IoT and Machine Learning for Smart Applications

This book provides an illustration of the various methods and structures that are utilized in machine learning to make use of data that is generated by IoT devices. Numerous industries utilize machine learning, specifically machine learning-as-a-service (MLaaS), to realize IoT to its full potential. On the application of machine learning to smart IoT applications, it becomes easier to observe, methodically analyze, and process a large amount of data to be used in various fields.

Features:

- Explains the current methods and algorithms used in machine learning and IoT knowledge discovery for smart applications
- Covers machine-learning approaches that address the difficulties posed by IoT-generated data for smart applications
- Describes how various methods are used to extract higher-level information from IoT-generated data
- Presents the latest technologies and research findings on IoT for smart applications
- Focuses on how machine learning algorithms are used in various real-world smart applications and engineering problems

It is a ready reference for researchers and practitioners in the field of information technology who are interested in the IoT and Machine Learning fields.

Designed cover image: ShutterStock

First edition published 2025
by CRC Press
2385 NW Executive Center Drive, Suite 320, Boca Raton FL 33431

and by CRC Press
4 Park Square, Milton Park, Abingdon, Oxon, OX14 4RN

CRC Press is an imprint of Taylor & Francis Group, LLC

ISBN: 9781032621081 (hbk)
ISBN: 9781032623269 (pbk)
ISBN: 9781032623276 (ebk)

DOI: 10.1201/9781032623276

Typeset in Times
by Newgen Publishing UK

IoT and Machine Learning for Smart Applications

Edited by
G. Vennira Selvi, T. Ganesh Kumar, M. Prasad,
Raju Hajare, and Priti Rishi

CRC Press
Taylor & Francis Group
Boca Raton London New York

CRC Press is an imprint of the
Taylor & Francis Group, an **informa** business

A CHAPMAN & HALL BOOK

Contents

About the Editors vii

Contributors ix

Chapter 1 Recent Advances in Machine Learning Strategies and
Its Applications.. 1

S. Saranyadevi and Lekshmi Gangadhar

Chapter 2 Understanding the Concept of IoT 19

*C. Emilin Shyni, Megha Menon K., and
Anesh D. Sundar ArchVictor*

Chapter 3 Unlocking the Power of IoT: An In-Depth Exploration 32

*Prakash Shanmurthy, Damodharan D.,
Anandhan Karunanithi, Pajany M., and J. John Bennet*

Chapter 4 Machine Learning in Internet of Things.............................. 67

Sujni Paul and Grasha Jacob

Chapter 5 Role of Machine Learning in Real-Life Environment 81

*Balasaraswathi V.R., Nathezhtha T., Vaissnavie V.,
and Rajashree*

Chapter 6 Efficient Blockchain-Based Edge Computing System
Using Transfer Learning Model 97

*P. Sivakumar, S. Nagendra Prabhu, S.K. Somasundaram,
J. Uma Maheswari, and Murali Murugan*

Chapter 7 Introducing a Compact and High-Speed Machine
Learning Accelerator for IoT-Enabled Health
Monitoring Systems 110

*Kavitha V.P., Magesh V., Theivanathan G., and
S.S. Saravana Kumar*

Chapter 8   Realization of Smart City Based on IoT and AI.............................. 127

*Pavithra N., Sapna R., Preethi, Manasa C.M., and
Raghavendra M. Devadas*

Chapter 9   Sentiment Analysis of Airline Tweets Using Machine Learning
Algorithms and Regular Expression... 143

*S. Nagendra Prabhu, A.P. Rohith, Shubhankar Bhope,
and P. Sivakumar*

Chapter 10 Smart Workspace Automation: Harnessing IoT and AI for
Sustainable Urban Development and Improved Quality of Life 163

*Sapna R., Preethi, Pavithra N., Manasa C.M., and
Raghavendra M. Devadas*

Chapter 11 Application of Digital Image Watermarking in the
Internet of Things and Machine Learning 178

K. Prabha and I. Shatheesh Sam

Index.. 197

About the Editors

G. Vennira Selvi is Professor at the School of Computer Science and Engineering and Information Science at Presidency University, Bengaluru, India. She received her Doctorate Degree in Computer Science and Engineering from Pondicherry University, Pondicherry in 2018 and M.E. degree in Computer Science and Engineering from Manonmaniam Sundaranar University, Tirunelveli. She has published 20 papers in reputed international journals, 6 papers as part of international conferences, and 12 papers as part of national conferences. She has published two books and nine book chapters. She has received eight Indian patents, one Design grant patent, and one international patent. She has 22 years of teaching experience in undergraduate and postgraduate courses. Her current research areas are IoT, wireless sensor networks, machine learning, and artificial intelligence.

T. Ganesh Kumar works as Associate Professor in the School of Computing Science and Engineering at Galgotias University, NCR, Delhi. He received an M.E. degree in Computer Science and Engineering from Manonmaniam Sundaranar University, Tamil Nadu, India. He completed his full-time Ph.D. degree in Computer Science and Engineering at Manonmaniam Sundaranar University. He was Co-Investigator for two Government of India-sponsored funded projects. He has publications in many reputed international Science Citation Index and Scopus indexed journals and conference proceedings. He has published more than ten Indian patents.

M. Prasad received his B.Tech, M.Tech, and doctorate degrees in Information Technology, Information Security, and Mobile Communication from Pondicherry University, Pondicherry, in 2008, 2010, and 2017, respectively. He is currently working in VIT Chennai campus School of Computing Science and Engineering. His current areas of research include IoT, security in mobile communication, cyber physical systems, machine learning, and artificial intelligence. He is having various professional memberships like IEEE, ACM, CSI, and Internet Society of India. He has published in more than 20 journals and holds 10 patents to his name.

Raju Hajare is Professor at the BMS Institute of Technology and Management. He is a doctorate in the field of "Nano Devices Modeling and Simulation". He has 19 years of academic and 2 years of industry experience at various levels of organizations. Dr. Raju has published more than 30 quality research publications in reputed Scopus-indexed international journals and IEEE proceedings. Professor Raju has two textbooks to his credit in the domain of electronics.

Priti Rishi is Associate Professor at the College of Engineering and Technology, SRM Institute of Engineering and Technology, Vadapalani Campus, Chennai. She received an M.E. degree in Electronics and Communication Engineering from Thapar Institute of Engineering and Technology, Punjab, India. She completed her part-time Ph.D. degree at the Faculty of Information and Communication Engineering at Anna University Chennai. She has multiple publications to her credit and holds more than three Indian patents.

Contributors

Anesh D. Sundar ArchVictor
Hodos Institute
Kirkland, Washington, USA

J. John Bennet
School of Computer Science and
    Engineering and Information
    Science
Presidency University
Bengaluru, India

Shubhankar Bhope
Department of Computational
    Intelligence
School of Computing
SRM Institute of Science and
    Technology
Chennai, India

Manasa C.M.
Department of Computer Science &
    Engineering
Presidency University
Bengaluru, India

Damodharan D.
Department of Computer Science and
    Engineering
Galgotias University
Greater Noida, India

Raghavendra M. Devadas
Department of Information
    Technology
Manipal Institute of Technology
Bengaluru, India
Manipal Academy of Higher Education
Manipal, India

Theivanathan G.
Velammal Engineering College
Tamil Nadu, India

Lekshmi Gangadhar
Nanodot Research Pvt. Ltd.
Trivandrum, India

Grasha Jacob
Government Arts and Science College
Tamil Nadu, India

Megha Menon K.
Department of Computer Science and
    Engineering Reva University
Bengaluru, India

Anandhan Karunanithi
Department of Computer Science and
    Engineering
GITAM School of Technology
GITAM University
Bengaluru, India

S.S. Saravana Kumar
Srinivasa Institute of Technology &
    Science
Jawaharlal Nehru Technological
    University Kakinada
Andhra Pradesh, India

Pajany M.
School of Computer Science and
    Engineering and Information Science
Presidency University
Bengaluru, India

J. Uma Maheswari
School of Computer Science and
   Engineering
Vellore Institute of Technology
Chennai, India

Murali Murugan
Macy's Inc.
Georgia, USA

Pavithra N.
Department of Computer Science &
   Engineering
Manipal Institute of Technology
Bengaluru, India
Manipal Academy of Higher Education
Manipal, India

Sujni Paul
CIS Division
Dubai Men's Campus
Higher Colleges of Technology
Dubai Academic City, UAE

K. Prabha
Department of Computer Science
Arunachala Arts and Science (Women)
   College
Tamil Nadu, India

S. Nagendra Prabhu
Department of Computational
   Intelligence
School of Computing
SRM Institute of Science and
   Technology
Chennai, India

Preethi
Department of Information Technology
Manipal Institute of Technology
Bengaluru, India
Manipal Academy of Higher Education
Manipal, India

Sapna R.
Department of Information Technology
Manipal Institute of Technology
Bengaluru, India
Manipal Academy of Higher Education
Manipal, India

Rajashree
Vellore Institute of Technology
Chennai, India

A.P. Rohith
Department of Computational
   Intelligence
School of Computing
SRM Institute of Science and
   Technology
Chennai, India

I. Shatheesh Sam
Department of PG Computer Science
Nesamony Memorial Christian College
Tamil Nadu, India

S. Saranyadevi
Department of Biotechnology
Paavai Engineering College
   (Autonomous)
Namakkal, India

Prakash Shanmurthy
School of Computer Science and
   Engineering and Information Science
Presidency University
Bengaluru, India

C. Emilin Shyni
School of Computer Science and
   Engineering and Information
   SciencePresidency University
Bengaluru, India

P. Sivakumar
School of Computer Science and
  Engineering
Vellore Institute of Technology
Chennai, India

S.K. Somasundaram
School of Computer Science and
  Engineering
Vellore Institute of Technology
Vellore, India

Nathezhtha T.
Vellore Institute of Technology
Chennai, India

Vaissnavie V.
Sathyabama Institute of Science and
  Technology
Chennai, India

Kavitha V.P.
SRM Institute of Science and
  Technology
Vadapalani Campus
Chennai, India

Balasaraswathi V.R.
Vellore Institute of Technology
Chennai, India

1 Recent Advances in Machine Learning Strategies and Its Applications

S. Saranyadevi and Lekshmi Gangadhar

1.1 INTRODUCTION

The domain of computer science called machine learning (ML) aids machines in learning without explicitly predetermined [1]. Rather than utilizing programing to perform tasks, ML comprises the conception of algorithms. Machines pick up knowledge from historical patterns and examples from their previous experience. Models are used to estimate new values can be constructed. When the amount of data and queries is too great for natural solutions, ML can be used to help identify answers by analyzing the data. It can make it easier for users to find crucial information faster. Because machines learn faster and can even outclass humans in a few sectors, they can easily tackle complex issues [2]. Its demand is constantly rising as a result. ML is growing in relevance along with big data and cloud computing because of how many problems it can solve with its computational capacity. It has a wider variety of uses in other fields, particularly because it mainly helps in the unearthing of novel drugs and enables physicians to diagnose patients accurately, enabling the early detection of numerous ailments.

In the field of data analysis and computing, artificial intelligence (AI), particularly ML, can be expanded quickly nowadays, usually enabling the applications to perform intelligently [3]. In general, ML is the well-liked new technology in the fourth industrial revolution (Industry 4.0) since it makes systems capable of automatically learning from experience and improving without the need for special programing [4–6].

1.2 MACHINE LEARNING TYPES

There are various categories of learning based on a range of potential circumstances regarding the accessibility of exams, training, and instructional materials, as outlined in the following sections.

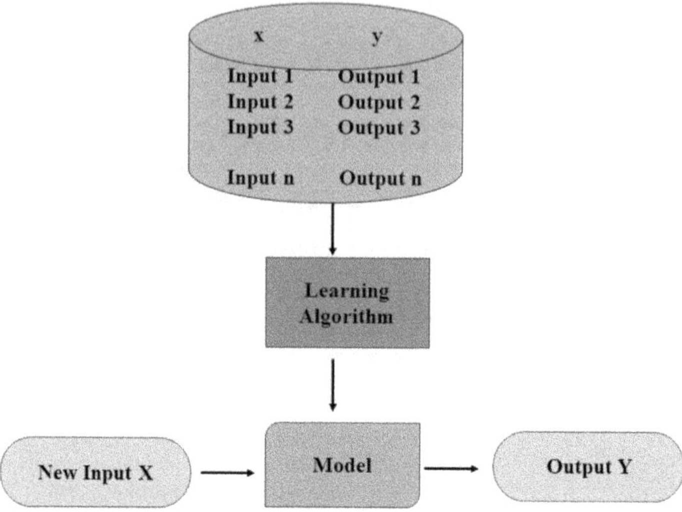

FIGURE 1.1   Supervised learning.

1.2.1 SUPERVISED LEARNING

One of the most extensively utilized learning techniques is supervised learning (SL), which involves feeding a data set as input and knowing the outputs for each matching input. As seen in Figure 1.1, the ML model utilizes them to attempt to develop a relationship between the feed and the result. There are two types of SL methods: "classification" and "regression" problems. The ML model translates the incessant output function toward the input variables in a regression issue. For example, the model needs to estimate a person's age based on an image of the person. The ML model attempts to map input variables to discrete types when it comes to classification problems. For instance, estimating home values according to location [7].

1.2.2 UNSUPERVISED LEARNING

We have an edge over all other learning algorithms with unsupervised learning (USL) since it attempts to solve problems whose outcomes and the influence of factors are unknown to us. As exemplified in Figure 1.2, the structures are found by grouping the provided data according to the correlation between the data's variables. In this case, the ML model's primary job is to group unsorted data according to similarities, patterns, and variations without requiring prior knowledge of the training set. The machine is restricted in its ability to identify the veiled structure in unlabeled data on its own. USL issues are categorized into: Clustering and association [7].

Clustering: In a clustering issue, we arrange the data points based on the connections between the variables. For example, we can group customers based on how they buy things from the stores. This approach is commonly employed for statistical examination of data across several domains. *Association*: In an association

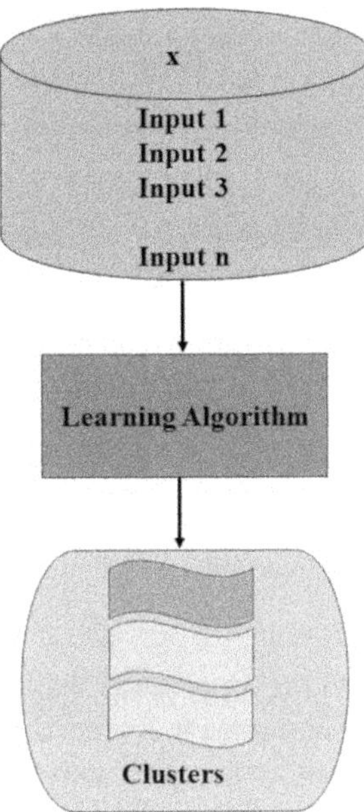

FIGURE 1.2 Unsupervised learning.

problem, we find patterns that characterize a significant section of the dataset, like predicting that buyers of X will probably purchase Y.

1.2.3 Semi-supervised Learning

An algorithm for semi-supervised learning (SSL) combines SL and USL. In ML sectors where they have unlabeled data and getting the labeled data from this is a laborious procedure, but it can be quite helpful [7]. We have a massive amount of input data in SSL Problems; a small portion is labeled, and the remainder is left unaltered. For instance, consider a photo library where only a few pictures (such as those of an elephant, automobile, or bridge) have labels; the remaining pictures are all unlabeled (Figure 1.3).

1.2.4 Reinforcement Learning

Reinforcement learning (RL) acts as a technique that produces an increasing number of positive outcomes. The response that the atmosphere delivers to the learning algorithm after it selects an output for a given input tells the algorithm how well the output lights the learner's objectives. Trial and error search and deferred outcomes are the two major components that regulate whether RL occurring [8]. Another major component in reinforcement learning (RL) is the **reward signal (R)**. The reward signal provides feedback on the agent's actions in terms of how favorable or unfavorable they are in achieving the desired outcome. It serves as a measure of success or failure,

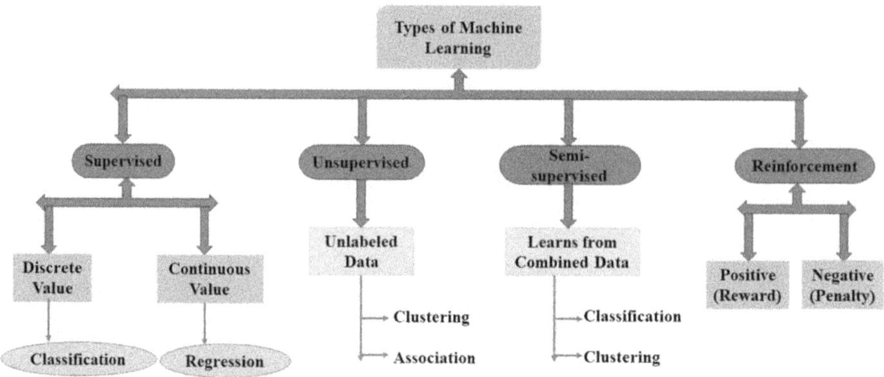

FIGURE 1.3   Strategies of machine learning.

encouraging behaviors that yield higher rewards while discouraging those leading to lower or negative rewards. This component is crucial in shaping the agent's behavior over time, as it learns to optimize actions that maximize cumulative rewards in the long run.

1.2.5   MULTITASK LEARNING

The primary aim of multitask learning is to assist other learners to help them advance and accomplish better. When this method is used on a given job, it essentially retains the memory of the steps involved in solving that specific problem as well as the way the learning algorithm responds to it to draw that conclusion. Similar techniques are employed by the algorithm to solve other comparable tasks. By allowing all students to share their experiences, the learning algorithm can be improved and executed more successfully, allowing for simultaneous and efficient learning [9].

1.2.6   ENSEMBLE LEARNING

In ensemble learning (EL), many separate learners are employed to create a single learner. This separate learner could be a decision tree, neural network, or Naïve Bayes. Since the 1990s, ensemble learning has gained popularity. When it comes to completing a particular activity, a group of learners is always preferable to an individual [10].

1.2.7   NEURAL NETWORK LEARNING

The biological reality of neurons, which are structures inside our brains that resemble cells, is where essentially the neural network starts. Understanding how neurons work is necessary for understanding brain networks. A neuron comprises four parts: The dendrites, the nucleus, the soma, and the axon. The electrical signals are picked up by the dendrites and transmitted to soma for processing.

This mechanism results in a drift to the dendritic. Terminals that guide the output to the subsequently linked neuron with the aid of an axon. The neural network that allows electrical signals to flow throughout the brain comprises these connections between neurons. An artificial neural network (ANN) functions similarly and has three layers [11]. Three primary forms of ANNs exist: Reinforcement, unsupervised, and supervised [12].

1.2.8 INSTANCE-BASED LEARNING

With this kind of learning, the student is sufficiently equipped to be capable of picking up patterns that attempt to apply to the material that it is fed. Thus, it goes by this name. The same lazy learner waits for the test set to reach before processing the training dataset with it. The drawback is that as data volume increases, so does the complexity of the system.

1.3 STATE OF THE ART

1.3.1 REAL-WORLD PROBLEMS

Computer-based systems can leverage all client information via ML. It responds to commands from the program and adjusts to novel circumstances or modifications.

Algorithms display previously unprogrammed characteristics and adjust to the data. A digital associate can scan emails and extract important ideas by learning to read and recognize context. This kind of learning involves having the ability to predict future consumer behavior. It enables you to be proactive as opposed to reactive and to have a deeper insight into your customers. ML can grow over time and applies to many different sectors and industries. The real-world uses of ML are exemplified in Figure 1.4.

1.3.2 CYBERSECURITY

Internet is prevalently employed for information and services alike. According to the author [13], nearly 48% of the world's populace has utilized the Internet as a source of acquaintance since 2017. In developed nations, this percentage rises to 82%, as the study [13] concludes. The Internet is a connectivity of various devices, networks, and computers. Its primary function is to transfer data between devices via networks. The improvements and breakthroughs in computer systems and mobile device networks caused a surge in Internet usage. Since most people use the Internet to obtain data, cybercriminals are more likely to target it [13]. It is considered stable when a computer-based system delivers idea integrity, obtainability, and secrecy. According to the study [14], if someone gains illegal access to the network to interfere with consistent operations, the integrity and security of the computer-based system would be jeopardized. Cybersecurity (CS), user assets, and cyberspace can be dwindling from unauthorized discrete outbreaks and access. According to a study by Thomas et al. [15], maintaining data availability, integrity, and confidentiality is the major objective of CS.

FIGURE 1.4   Real-time uses of machine learning.

1.3.3 Healthcare

Many changes are occurring in the domains of industry, transportation, and government because of developments in the area of deep learning (DL) and ML. Over the decades, many studies have been conducted on DL. DL has been used extensively and has produced state-of-the-art results in several areas, like computer vision, text analytics, and speech processing [16]. Recently, DL/ML strategies have been executed in the healthcare field, and researchers have shown remarkable results in tasks, such as brain tumor division, medical image reconstruction, lung nodule identification, types of lung disease, the identity of body parts, etc. [17].

CAD systems might deliver a second view that will help radiologists in corroborating the diagnosis. DL and ML will progress the functionality of CAD and other systems that will help radiologists in making judgments. Technological advancements in big data, cloud-computing, edge-computing, and mobile connectivity also facilitate the usage of ML and DL algorithms in the field of healthcare [18]. Together, they can increase prediction accuracy and promote the development of an intelligent, human-centered solution.

1.3.4 Intelligent Transportation System

Intelligent transportation is an outcome of implementing communication, information, and sensing technology in transit and transportation systems. Intelligent transportation systems, which include driverless cars, public transportation system-based management, and traveler information systems followed by road traffic management, are essential components of smart cities. These services are anticipated to have a significant positive impact on society by reducing contamination, increasing energy effectiveness, and improving transit and transportation efficacy, which ultimately improves traffic and road safety. Using techniques like RL and DL, which have gained popularity recently in DL and ML models, allows for the correct generation of decisions and predictions by utilizing patterns [19].

1.3.5 Renewable Energy

A growing number of people are looking for eco-friendly and other forms of energy due to the detrimental impacts of burning fossil fuels and the fast depletion of these resources. As stated in this chapter, the energy industry is expanding its usage of renewable energy sources, including biomass, wind, solar thermal, tidal waves, geothermal, and solar photovoltaic. There will be variability in the electrical networks for several reasons, such as when supply exceeds demand or when demand exceeds supply. ML is applied to energy management and optimization [20]. Table 1.1 exemplifies the different kinds of ML strategies with examples.

1.3.6 Smart Manufacturing and Grid

Various types are developed for manufacturing, and it is termed "Smart Manufacturing", which uses digital technology, worker training, fast design modifications, and high adaptability in its manufacturing process. Efficient production recycling, optimization

TABLE 1.1
Different ML Strategies with Examples

S. No.	Examples	Model Development	Types of Learning	Strategies
1	Regression and Classification	Models that utilize labeled data for learning	Supervised	Task-driven
2	Reduction of dimensionality, association, and clustering	Models that utilize unlabeled data for learning	Unsupervised	Data-driven
3	Category and clustering	By utilizing joint data models were developed by employing the concept of penalty and reward as a base for model development	Semi supervised	Labeled and unlabeled
4	Category and control		Reinforcement	Environment-driven

of supply chains, and demand-driven, fast adjustments to production levels are among the other duties. Smart manufacturing is made possible by advancements in robotics, massively parallel processing power, and industry-wide networking for gadgets and services. Despite criticism of the electrical energy grid's basic architecture, this design has endured consistently over time. The electrical system remains unchanged over the centuries, and we are living in the 21st century. But power has become more and more necessary as population and consumption have increased [21].

1.4 PROBLEMS

 i. Demand analysis is challenging
 ii. Response times are sluggish

An innovative smart grid concept has emerged to report the problems. SG is a vast energy network that uses intelligent and real-time monitoring, communication, control, and self-healing technologies to ensure the reliability and security of clients' electricity supply while offering a range of options. SGs are advanced cyber-physical systems. The modern SG's functionality can be divided into four categories.

1. *Consumption*: Energy is utilized by various industries and people for numerous reasons.
2. *Distribution*: Power is disseminated to enable broader use.
3. *Transmission*: Electricity is transferred via a higher-voltage electronic infrastructure.
4. *Generation*: Energy is generated using various techniques during this stage.

In the area of SG, ML, and DL functionalities have forecasting fraud detection, optimal scheduling, stability of the SG, identification of security breaches, identification of network anomalies, sizing, identification of faults, energy consumption, price, and energy generation.

1.4.1 Computer Network

Important technical developments in networking, like network programmability through Software-Defined Networking (SDN), have contributed to the utility of ML in networking. Although ML has been applied widely to report problems, including speech, and pattern recognition, its application in network processes and management has been restricted. The main challenges are guessing what kind of data can be assembled and what kind of controls can be executed on older network hardware. These problems are mitigated by the SDN's capacity to program the network. Routine network administration and operation tasks are automated with the use of ML-based cognition. It is, therefore, fascinating and problematic to apply ML methods to such wide and complex networking problems. Because of this, ML in networking is an intriguing field of study that calls for a deep comprehension of both networking and ML methodologies.

1.4.2 ENERGY SYSTEM

A group of well-organized parts used for managing energy, manufacturing, and/or transformation is called an energy system [22]. Mechanical, thermal, chemical, and electro-magnetically elements might be united to form energy systems that cover a broad range of energy-related categories, encompassing renewables and other types of energy [22]. An enormous quantity of data grouped results in the increasing use of data-gathering devices in power systems. Energy system development must make difficult decisions to satisfy a range of exacting and competing objectives, including budgetary constraints, ecological impact, effectiveness, and operational efficiency. Smart sensors are progressively being used in the manufacture and utilization of energy [23].

Making educated decisions is more problematic due to the abundance of opportunities presented by big data. The execution of big data technology in many uses has been facilitated by using ML models. The energy industry has seen a rise in the usage of prediction techniques based on ML models because they facilitate the inference of functional correlations from observations [24]. ML models in energy systems are essential for demand, making, and consumption predictive modeling because of their speed, accuracy, and effectiveness. ML models provide an understanding of how the energy system functions in the setting of intricate human relationships [25]. Both conventional energy systems and substitute and renewable energy systems are created using ML algorithms.

1.5 RECENT ADVANCES IN REAL-TIME USES

1.5.1 ML FOR ITS

Rapid car proliferation, inadequate transportation infrastructure, and the nonexistence of road security regulations are causing serious ecological and quality-of-life issues in urban areas. For example, huge trucks frequently break the rules of conventional highways in metropolitan areas, causing traffic jams and delays. Furthermore, many bikers have frequented near-misses because of clothing, posture changes, partial occlusions, and various observations that pointedly hinder the finding rates of ML strategies. Enhancing the identification and classification of pedestrians, cyclists, and special vehicles and license plate recognition (LPR) for a more sustainable and safer has garnered significant attention in the past ten years because of the usage of machine learning (ML) and deep learning (DL) strategies for the evaluation and presentation of a vast number of data collected from various sources. While a great range of appearances can be captured by deep models, environment adaptation is crucial. DL success is based on ANNs that mimic images by incorporating interrelated node system groups that behave like the human brain. The nodes in a nearby layer will be comprised of connections whose weights originate from nodes in another layer. An instigation function in a node receives an input and a weight to generate an output value.

1.5.2 ML FOR HEALTHCARE

Activities and annotations are provided as input to policy functions throughout actions carried out as response rewards. The technique acquired from this policy function is called RL. RL is executed for various healthcare-based uses. Even RL is pragmatic to the widespread symptom-checking method of disease detection. GOGAME is another possible usage of RL [26].

When there are vast numbers of unlabeled and fewer labeled information accessible, SSL is appropriate. SSL uses both unlabeled and labeled data for training. Applications for SSL in healthcare include medical picture segmentation [27], activity recognition utilizing a variety of sensors suggested, and SSL for medical care data clustering employed by the author [27]. Typical uses of SL in the medical domain include nodule identification in pulmonary imaging and the recognition of human parts utilizing several picture modalities.

USL involves training the model with unlabeled data to map input to output: Clustering is based on (i) similarity, (ii) feature selection and dimensionality reduction, and (iii) finding anomalies [28]. Various healthcare uses, including feature selection by Principal Component Analysis and heart disorder estimation using clustering, can benefit from USL. Prognosis, finding, treatment, and medical workflow are the four major fields of healthcare that can turnover from ML or DL techniques.

1.5.3 ML FOR CYBERSECURITY

In several industries like CS, design and manufacturing, medical [29], education, and finance, AI and ML are extensively recognized and utilized. The subsequent domains of CS – intrusion identification, dark or deeper websites, phishing, malware recognition, fraud, and spam identification – all make extensive use of ML techniques. Vigorous and innovative approaches are required to tackle the problems of CS as times change. Given that ML is experience-based, it is amenable to evolutionary attacks. Classical CS technologies include firewalls, interruption preclusion systems, antivirus software, unified threat management [30], and Security Information and Event Management solutions.

The author concluded that AI-based CS systems outperform classical CS systems in terms of mistake rate, performance, and post-cyberattack response. According to the study, cyberspace harm caused by an assault is only detected after it has occurred, and this happens in about 60% of cases [31]. ML has a higher hold on both the attacker and CS sides. Regarding CS, the goal is to defend everything from harm created by attackers, identify attacks early on, and ultimately optimize performance. On the invader's side, ML is utilized to recognize holes and system vulnerabilities and methods to become past firewalls. The study comes to the same conclusion regarding how to improve classification performance by ML strategies.

1.5.4 ML FOR COMPUTER NETWORKS

1.5.4.1 Traffic Prediction

Traffic forecasting is crucial to the effective management and execution of network operations since networks are growing more diverse and complicated every day, making it harder to administer and carry out network processes. Time series estimating projects traffic patterns for the near future. Classifying network traffic is a crucial step in managing and carrying out network operations. This process involves resource provisioning, performance monitoring, and differentiation of services.

1.5.4.2 Congestion Control

Overflowing packets in a system will be regulated by a congestion controller. It ensures that resource use is fair, network stability is maintained, and the packet loss ratio stays within a satisfactory range.

1.6 EXISTING CHALLENGES TO ML TECHNOLOGY

ML is not without its tests, while technology shows potential and already serves businesses worldwide. ML, for example, is good at identifying patterns but not so good at generalizing information. Additionally, there is the problem of users becoming weary of algorithms. In ML, a reasonable quantity of data and resources with good performance are required for model training. Several Graphics Processing Units are used to tackle this task. Real-time engineering uses require ML techniques that are modeled to robustly solve a specific problem. It is essential to create a model specifically for every task in real-time use since a single model created for one activity usage cannot handle all the tasks across multiple domains.

Within the field of medicine, ML has the ability to anticipate illness incidence and detect terrorist acts. As in Ribeiro et al. [32], relying solely on ML forecasts will not prevent disastrous outcomes. While ML strategies are employed in many domains, they demand extremely high degrees of correctness in certain domains as a substitute for accuracy and speed. For a model to be considered trustworthy, it must be assured that there is no shift in the dataset during training and testing. This was performed by preventing data leaks [33].

The location of a moving object can be determined by permitting technologies, including GPS and mobile phones; a crucial problem for ML is keeping this data safe and unchangeable. The author claims that an object's position data from numerous sources are compared to identify similarities; the study corroborates the ambiguity in the position data composed from numerous sources due to network delays and that the veracity of such data desires to be addressed by ML strategies.

To facilitate trustworthy interactions among consumers and service providers in a networked web system, an ontology of trust is put out [34]. Text classification also makes use of trustworthiness. Trustworthiness combines both semantic and practical relations when interpreting the text's meaning. It uses a metric model to verify the software's reliability. Like this study [35,36], the author claims that by using ML strategies to create power-aware strategies, power usage in businesses and data centers can be reduced. It's advisable to try the machines off energetically to cut the usage completely. The forecasting model will determine which machine should be turned off; it is crucial to have faith in this model.

Certain problems associated with ML have important effects that are currently being felt. The first is the "black box issue", or the lack of interpretability and explainability. Even the people who created it are baffled by how ML models come to make their own decisions and actions. This makes it challenging to fix errors and guarantee the impartiality and precision of a model's output. For example, when it was identified that women were given significantly smaller credit lines by Apple's credit card model than men, the company was not able to provide an explanation or fix the issue. This relates to data and algorithmic bias, which are the two biggest issues plaguing the area. Algorithms are regularly prejudiced against women, Americans, and people of various civilizations. The topmost AI research center in the world, Google DeepMind, has warned that LGBT people may be at risk from the technology.

Even though this problem is widespread and well-known, there was opposition to the significant action that several industry professionals believe is required. Researchers, legislators, and activists were surveyed, and the majority voiced concern that as AI advances by 2030. The entire country is currently discussing and passing laws about AI, especially regarding applications that are instantly and harmful recognition for police enforcement.

1.7 UTILIZATIONS OF ML

In Industry 4.0, ML strategies are more common because of the capacity for intelligent decision-making and for learning from the past. Here, we go over and condense several application areas for ML algorithms.

1.7.1 INTELLIGENT DECISION-MAKING

ML strategies are applied to create intelligent decisions using data-driven [37]. The foundation of predictive analytics is the capability to predict unknown results by utilizing past events and identifying and catching the relationship between the forecast and descriptive variables like credit card scams and criminal findings followed by the crime. Prognostic analytics and intellectual decision-making are useful in the retail sector for preventing out-of-stock situations, managing inventories, better understanding customer behavior and preferences, and streamlining transportation and warehousing. The most often utilized methods in the fields are ANN, decision trees, and support vector machines [37]. Precisely forecasting the result can benefit any firm, including those in social interaction, transport, sales and advertising, healthcare, banking, financial facilities, telecommunication, E-commerce, and other businesses.

1.7.2 CYBER-SECURITY AND HAZARD INTELLIGENCE

Cybersecurity (CS) is a key component of Industry 4.0 and is responsible for safeguarding data, hardware, systems, and networks. One of the most significant technologies in CS is ML, which protects users by encrypting cloud data, anticipating malicious users online, identifying internal threats, and detecting malware in traffic. CS and threat intelligence make use of DL-based security models, association rule learning approaches, and ML categorization models [38].

1.7.3 SMART CITIES

By giving items the ability to send information and carry out tasks without the necessity for human interaction, all objects in the Internet of Things (IoT) are transformed into things. Business, healthcare, farming, retail, transportation, communiqué, edification, smart homes, smart governance, and smart cities [39] are a few industries that use IoT. Because ML can evaluate data and forecast future events, it has emerged as a key tool in the IoT. For example, in smart cities, congestion can be forecast, judgments depend on the environment, energy can be estimated for a specific time, and parking availability can be projected.

1.7.4 SUSTAINABLE AGRICULTURE

All human endeavors depend on farming to thrive [40]. Sustainable farming practices enhance crop yield while reducing adverse environmental effects. The authors [41,42] clarify how novel technologies, mobile devices, and the IoT can be utilized to collect vast amounts of information, which in turn can support the adoption of sustainable agriculture techniques. Sustainable agriculture develops knowledge-intensive supply chains using technology, skills, and information. The processing, production, and preproduction phases of agriculture, as well as the distribution phases, can all benefit from the application of various ML techniques. These include consumer study, inventory control, production scheduling, estimating animal needs, managing soil nutrients, identifying weeds and diseases, forecasting the weather, determining irrigation needs, analyzing soil characteristics, and projecting crop yields.

1.7.5 COVID-19 PANDEMIC AND HEALTHCARE

ML may help address diagnostic and prognostic challenges in a range of medically relevant application fields, such as sickness prediction, medical information extraction, data regularity detection, patient data management, etc. [43]. The WHO classifies coronavirus as an infectious disease. In the fight against COVID-19, learning strategies have recently gained prominence. The COVID-19 pandemic's death rate, high-risk patients, and other anomalies are being categorized using learning techniques. It is utilized to its fullest to recognize the source of the virus, forecast the COVID-19 outbreak, and identify and manage the illness [44]. ML may be used by researchers to forecast the locations and times at which COVID-19 would spread, allowing those areas to be prepared. DL can offer more effective solutions to the therapeutic image processing issues related to the COVID-19 epidemic. All things considered, DL and ML processes can aid in fighting the coronavirus and epidemic, possibly even helping create smart clinical decisions in the medical field.

1.7.6 PATTERN RECOGNITION

The detection of objects in images is the goal of image recognition, which makes extensive use of ML [45]. A few examples of image recognition are the cancer label on an X-ray image, face detection, character recognition, and social media suggestion tagging. The well-known language and sound models in speech recognition are Alexa, Siri, Cortana, Google Assistant, etc. Pattern recognition is the automatic identification of regularities and patterns in data, like image study [45]. Various ML strategies are used in this field, like feature selection, classification, clustering, and sequence labeling.

1.8 TRIALS AND FUTURE GUIDELINES IN ML

This chapter examines the pertinence of various ML algorithms in the study of intelligent data and applications, which raises several research concerns. Research prospects and possible future initiatives are outlined and explored here.

The overview of research directions:

i.   It's crucial to focus on the thorough examination of data-collecting methods while working with real-world data. It is important to enhance existing preprocessing techniques or create new ones to manage real-world data associated with usage domains.

ii. One of the research interests is determining which ML technique is best for the intended application.

iii. There is a great deal of interest in academia in proposing new hybrid algorithms and enhancing or modifying existing ML algorithms to make them more applicable to the domain of target applications.

The efficacy and efficiency of ML solutions are determined by the nature as well as the qualities of the information, along with the performance of ML processes over it. The development of vast volumes of data in a short period by numerous application domains, like cyber-security, healthcare, as well as agriculture, makes data collecting in these areas challenging. Relevant data collecting is essential to moving on with the investigation of the data in ML-based uses. Consequently, while working with real-world information, attention must be paid to a deeper analysis of the data collection techniques.

We must do the challenging process of purifying the acquired data from numerous sources because there might be numerous outliers, missing and confusing values in the prevailing information, which will impair the training of ML strategies. Therefore, to maximize the utilization of ML strategies, pretreatment should be enhanced, and a novel preprocessing procedure should be used.

The features and nature of the information might affect the results of the various ML strategies, making it difficult to choose the ML procedure that is better opted for the target use, for removing understandings, and for data analysis. An improper ML process will provide unforeseen results, which could lessen the model's efficacy and accuracy. These algorithms are adjusted for the intended usage domains, or novel techniques must be suggested.

The uses and their associated ML-based solutions' final success will be based on ML algorithms and the characteristics of the data. When the data is not representative, contains irrelevant characteristics, is of low quality, or is not enough for training, ML models will produce fewer accurate results and become unusable. Two important things are needed to create an intelligent application: Managing different learning strategies and efficiently processing data.

Many new questions in the field of ML methods for intelligent data and applications are brought up by our study. Consequently, we highlight the challenges discussed in this section together with potent avenues for upcoming study and activities. The efficacy and efficiency of an ML-based solution are based on the type and quality of the information as well as the functioning of the learning procedures. To compile data in a particular field, like IoT, CS, healthcare, farming, etc. Data is thus gathered for the intended ML-based applications. It is necessary to thoroughly analyze data-gathering procedures while working with real-world information. Moreover, there may be a significant amount of missing data, outliers, unclear values, and meaningless data in historical data.

Numerous ML algorithms are available to evaluate data and derive perceptions, but both the data and the learning procedures play a major part in the overall performance of an ML-based solution and its supporting applications. Reduced accuracy will result from data that was critical to learn from, like non-representative, low-quality, inappropriate features, or not enough of it for training. Because of this, it's essential to develop an ML-based solution, design intelligent apps, process data accurately, and manage different learning algorithms.

1.9 CONCLUSION

This study on ML techniques includes an extensive examination of data analysis. Here, a summary of real-world problems and how various learning processes are employed to produce answers is provided. The performance of ML procedures and the properties of data will determine the accomplishment of the ML model. ML algorithms must be educated by data collected from diverse real-world scenarios and familiar with target application knowledge to produce intelligent decision-making. In this overview, a range of application areas and challenges in the real world are highlighted about the applicability of ML techniques. Finally, a summary and discussion of future research prospects and other problems are provided. Every problem in the target uses field needs to be solved with the help of efficient solutions. This chapter will function as a sample point for experts in the industry as well as academia. From the methodological viewpoint, it also serves as a yardstick for decision-makers across various usage realms followed by real-world situations. The applications of ML are not limited to a single industry. Instead, it is affecting various areas, like gaming, broadcasting and entertainment, information technology, lending and finance, and the car industry. Because ML has so many applications, researchers are working in several fields to attempt and change the nation.

REFERENCES

1.   Sharma, D. and Kumar, N., 2017. A review on machine learning algorithms, tasks and applications. *International Journal of Advanced Research in Computer Engineering & Technology (IJARCET)*, 6(10), pp. 215–223.

2. Sandhu, T.H., 2018. Machine learning and natural language processing – a review. *International Journal of Advanced Research in Computer Science*, 9(2), pp. 582–584.

3. Sarker, I.H., Furhad, M.H. and Nowrozy, R., 2021. AI-driven cybersecurity: an overview, security intelligence modeling and research directions. *SN Computer Science*, 2, pp. 1–18.

4. Sarker, I.H., Hoque, M.M., Uddin, M.K. and Alsanoosy, T., 2021. Mobile data science and intelligent apps: concepts, AI-based modeling and research directions. *Mobile Networks and Applications*, 26, pp. 285–303.

5. Sarker, I.H., Kayes, A.S.M., Badsha, S., Alqahtani, H., Watters, P. and Ng, A., 2020. Cybersecurity data science: an overview from machine learning perspective. *Journal of Big Data*, 7, pp. 1–29.

6. Ślusarczyk, B., 2018. Industry 4.0 – are we ready? *Polish Journal of Management Studies*, 17(1), pp. 232–248.

7. Zhu, X. and Goldberg, A.B., 2020. *Introduction to semi-supervised learning* (Vol. 109, pp. 373–440). Cham, Switzerland: Springer Nature.

8. Sutton, R.S., 1992. Introduction: the challenge of reinforcement learning. In Richard S. Sutton (ed.), *Reinforcement learning* (pp. 1–3). Boston, MA: Springer US.

9. Caruana, R., 1997. Multitask learning. *Machine Learning*, 28, pp. 41–75.

10. Opitz, D. and Maclin, R., 1999. Popular ensemble methods: an empirical study. *Journal of Artificial Intelligence Research*, 11, pp. 169–198.

11. Sharma, V., Rai, S. and Dev, A., 2012. A comprehensive study of artificial neural networks. *International Journal of Advanced Research in Computer Science and Software Engineering*, 2(10), pp. 278–284.

12. Hiregoudar, S.B., Manjunath, K. and Patil, K.S., 2014. A survey: research summary on neural networks. *International Journal of Research in Engineering and Technology*, 3(15), pp. 385–389.

13. Bhattacharyya, D.K. and Kalita, J.K., 2013. *Network anomaly detection: A machine learning perspective*. Boca Raton, FL: CRC Press.

14. Ambalavanan, V., 2020. Cyber threats detection and mitigation using machine learning. In T. Thomas, A. P. Vijayaraghavan, and S. Emmanuel (eds.), *Handbook of research on machine and deep learning applications for cyber security* (pp. 132–149). Hershey, PA: IGI Global.

15. Thomas, T., Vijayaraghavan, A.P. and Emmanuel, S., 2020. *Machine learning approaches in cyber security analytics* (pp. 37–200). Singapore: Springer.

16. Latif, S., Qadir, J., Farooq, S. and Imran, M.A., 2017. How 5G wireless (and concomitant technologies) will revolutionize healthcare?. *Future Internet*, 9(4), p. 93.

17. Havaei, M., Davy, A., Warde-Farley, D., Biard, A., Courville, A., Bengio, Y., Pal, C., Jodoin, P.M. and Larochelle, H., 2017. Brain tumor segmentation with deep neural networks. *Medical Image Analysis*, 35, pp. 18–31.

18. Latif, S., Asim, M., Usman, M., Qadir, J. and Rana, R., 2018. Automating motion correction in multishot MRI using generative adversarial networks. *32nd Conference on Neural Information Processing Systems (NIPS 2018)*, Montréal, Canada. arXiv preprint arXiv:1811.09750.

19. Mao, Q., Hu, F. and Hao, Q., 2018. Deep learning for intelligent wireless networks: a comprehensive survey. *IEEE Communications Surveys & Tutorials*, 20(4), pp. 2595–2621.

20. Deshmukh, M.K. and Deshmukh, S.S., 2008. Modeling of hybrid renewable energy systems. *Renewable and Sustainable Energy Reviews*, 12(1), pp. 235–249.

21. Mosavi, A., Salimi, M., Faizollahzadeh Ardabili, S., Rabczuk, T., Shamshirband, S. and Varkonyi-Koczy, A.R., 2019. State of the art of machine learning models in energy systems, a systematic review. *Energies*, 12(7), p. 1301.

22. Smets, A., Jäger, K., Isabella, O., Van Swaaij, R. and Zeman, M., 2016. *Solar energy: the physics and engineering of photovoltaic conversion, technologies and systems*. London: Bloomsbury Publishing.

23. Torabi, M., Hashemi, S., Saybani, M.R., Shamshirband, S. and Mosavi, A., 2019. A Hybrid clustering and classification technique for forecasting short-term energy consumption. *Environmental Progress & Sustainable Energy*, 38(1), pp. 66–76.

24. Yildiz, B., Bilbao, J.I. and Sproul, A.B., 2017. A review and analysis of regression and machine learning models on commercial building electricity load forecasting. *Renewable and Sustainable Energy Reviews*, 73, pp. 1104–1122.

25. Abidi, M.H., Alkhalefah, H., Moiduddin, K., Alazab, M., Mohammed, M.K., Ameen, W. and Gadekallu, T.R., 2021. Optimal 5G network slicing using machine learning and deep learning concepts. *Computer Standards & Interfaces*, 76, p. 103518.

26. Kao, H.C., Tang, K.F. and Chang, E., 2018, April. Context-aware symptom checking for disease diagnosis using hierarchical reinforcement learning. In *Proceedings of the AAAI conference on artificial intelligence* (vol. 32, no. 1), pp. 1–7.

27. Sohail, M. N., Ren, J. and Uba Muhammad, M., 2019. A Euclidean group assessment on semi-supervised clustering for healthcare clinical implications based on real-life data. *International Journal of Environmental Research and Public Health*, 16(9), p. 1581.

28. Ramana, K., Ponnavaikko, M. and Subramanyam, A., 2019. A global dispatcher load balancing (GLDB) approach for a web server cluster. In *ICCCE 2018: proceedings of the international conference on communications and cyber physical engineering 2018* (pp. 341–357). Springer Singapore.

29. Jha, S. and Topol, E.J., 2016. Adapting to artificial intelligence: Radiologists and pathologists as information specialists. *JAMA*, 316(22), pp. 2353–2354.

30. Zhang, X. and Chow, K.P., 2020. A framework for dark web threat intelligence analysis. In *Cyber warfare and terrorism: concepts, methodologies, tools, and applications* (pp. 266–276). Hershey, PA: IGI Global.

31. Sharma, A., Kalbarczyk, Z., Barlow, J. and Iyer, R., 2011, June. Analysis of security data from a large computing organization. In *2011 IEEE/IFIP 41st international conference on dependable systems & networks (DSN)* (pp. 506–517). IEEE.

32. Ribeiro, M.T., Singh, S. and Guestrin, C., 2016, August. "Why should I trust you?" Explaining the predictions of any classifier. In *Proceedings of the 22nd ACM SIGKDD international conference on knowledge discovery and data mining,* San Francisco, California, pp. 1135–1144.

33. Ghosh, S., Lincoln, P., Tiwari, A. and Zhu, X., 2017, March. Trusted machine learning: model repair and data repair for probabilistic models. In *Workshops at the thirty-first AAAI conference on artificial intelligence,* San Francisco, California, pp. 143–152.

34. Zhu, M. and Jin, Z., 2009, September. A trust measurement mechanism for service agents. In *2009 IEEE/WIC/ACM international joint conference on web intelligence and intelligent agent technology* (vol. 1, pp. 375–382). New York, NY: IEEE.

35. Tao, H. and Chen, Y., 2009, September. A metric model for trustworthiness of softwares. In *2009 IEEE/WIC/ACM international joint conference on web intelligence and intelligent agent technology* (vol. 3, pp. 69–72). New York, NY: IEEE.

36. Berral, J.L., Goiri, Í., Nou, R., Julià, F., Guitart, J., Gavaldà, R. and Torres, J., 2010, April. Towards energy-aware scheduling in data centers using machine learning. In *Proceedings of the 1st international conference on energy-efficient computing and networking* (pp. 215–224). New York, NY: ACM (Association for Computing Machinery).

37. Li, X.H., Cao, C.C., Shi, Y., Bai, W., Gao, H., Qiu, L., Wang, C., Gao, Y., Zhang, S., Xue, X. and Chen, L., 2020. A survey of data-driven and knowledge-aware explainable AI. *IEEE Transactions on Knowledge and Data Engineering*, 34(1), pp. 29–49.

38. Sarker, I.H., Abushark, Y.B., Alsolami, F. and Khan, A.I., 2020. IntruDTree: a machine learning based cyber security intrusion detection model. *Symmetry*, 12(5), p. 754.

39. Zanella, A., Bui, N., Castellani, A., Vangelista, L. and Zorzi, M., 2014. Internet of things for smart cities. *IEEE Internet of Things Journal*, 1(1), pp. 22–32.

40. Adnan, N., Nordin, S.M., Rahman, I. and Noor, A., 2018. The effects of knowledge transfer on farmers decision making toward sustainable agriculture practices: In view of green fertilizer technology. *World Journal of Science, Technology and Sustainable Development*, 15(1), pp. 98–115.

41. Kamble, S.S., Gunasekaran, A. and Gawankar, S.A., 2018. Sustainable Industry 4.0 framework: A systematic literature review identifying the current trends and future perspectives. *Process Safety and Environmental Protection*, 117, pp. 408–425.

42. Kamble, S.S., Gunasekaran, A. and Gawankar, S.A., 2020. Achieving sustainable performance in a data-driven agriculture supply chain: a review for research and applications. *International Journal of Production Economics*, 219, pp. 179–194.

43. Ardabili, S.F., Mosavi, A., Ghamisi, P., Ferdinand, F., Varkonyi-Koczy, A.R., Reuter, U., Rabczuk, T. and Atkinson, P.M., 2020. COVID-19 outbreak prediction with machine learning. *Algorithms*, 13(10), p. 249.

44. Alakus, T.B. and Turkoglu, I., 2020. Comparison of deep learning approaches to predict COVID-19 infection. *Chaos, Solitons & Fractals*, 140, p. 110120.

45. Chiu, C.C., Sainath, T.N., Wu, Y., Prabhavalkar, R., Nguyen, P., Chen, Z., Kannan, A., Weiss, R.J., Rao, K., Gonina, E. and Jaitly, N., 2018, April. State-of-the-art speech recognition with sequence-to-sequence models. In *2018 IEEE international conference on acoustics, speech and signal processing (ICASSP)* (pp. 4774–4778). New York, NY: IEEE.

2 Understanding the Concept of IoT

C. Emilin Shyni, Megha Menon K.,
and Anesh D. Sundar ArchVictor

2.1 INTRODUCTION

Imagine waking up to the sound of your favorite melody, with the curtains in your bedroom gently pulling back to reveal the first rays of sunshine, your coffee machine already brewing the perfect cup to jumpstart your day. This isn't a scene from a futuristic movie; it's a slice of the present, brought to life by the Internet of Things (IoT). A silent revolution has unfurled throughout homes and cities around us, connecting devices and everyday objects to work seamlessly, making our lives more comfortable and efficient. But what exactly lies behind this technological evolution that's subtly yet persistently transforming our daily routines?

IoT is no longer a buzzword confined to tech enthusiast circles; it has become an integral part of our lives, and its relevance is proliferating with each passing day. This intricate web of interconnected gadgets and sensors offers a symphony of convenience that orchestrates our thermostats, vehicles, and even healthcare devices. It's a concept that weaves itself into the very fabric of our existence, aiming to optimize our world by enhancing communication between devices – as much as we humans communicate with each other.

Yet, despite its ubiquitous nature, the essence of IoT is shrouded with a sense of mystique for many.

For the uninitiated, the IoT can seem overwhelming – a complex network of intelligent objects turning the mundane into the magical. But fear not, for understanding IoT doesn't require a degree in rocket science; it simply demands curiosity and a willingness to embrace the future. As we usher in an era where your car notifies your heater to warm up the house in anticipation of your arrival, isn't it time to demystify the magic and grasp the strings that control this technological puppetry?

This blog post aims to pull back the curtain on IoT, revealing the wizardry of interconnected tech in a language you'll not only comprehend but relate to. We're about to dive into the sprawling ecosystem of the IoT to unveil how it shapes our modern lives and, more importantly, how it stands to revolutionize the future. Ready to embark on a journey of discovery that will change the way you interact with the world around you? Then let's get connected, for this is a narrative that stretches from the smallest microchip to the largest cities, narrating a story where you are an inseparable protagonist.

DOI: 10.1201/9781032623276-2

2.2 DEFINITION OF IOT

To put it simply, the IoT is a network of linked objects that are capable of exchanging data and interacting with one another. These gadgets range from medical equipment and animals to industrial gear, transit systems, cell phones, and home appliances.

Increasing the intelligence and connectivity of common things to enable data transmission and reception, remote control, and interaction with other systems and devices is the aim of the IoT. This has the potential to completely transform the way we work and live by enabling innovative and effective methods of operation and by giving governments and corporations access to insightful information.

2.3 HISTORICAL CONTEXT AND EVOLUTION OF IOT

Are you aware that the idea of the IoT has existed for some time? You might be shocked to hear that Kevin Ashton, the "father of IoT", originally used the word in 1998 while he was employed by Procter & Gamble. When smartphones and tablets became extensively used in 2011 and 2012, the IoT began to receive attention in the media. These days, Industry 4.0, smart cities, and smart homes are just a few areas in which the IoT is included. Part 1 of this two-part chapter will examine the evolution of IoT and its historical context.

Historical Context:
The concept of IoT was first coined in the early 1990s when researchers started exploring the possibility of connecting everyday objects to networks. However, the term "Internet of Things" was coined in 1998 by Kevin Ashton, a British entrepreneur who worked at Procter & Gamble.

Evolution of IoT:

1. Early Stage (1990s–2000s):
   In the early stages, IoT research focused on connecting everyday objects to networks. However, it wasn't until the early 2000s that the idea of IoT gained mainstream attention.
2. Growth Stage (2010–2012):
   The widespread adoption of smartphones and tablets led to the creation of IoT applications. Companies started developing IoT-enabled devices that could be controlled and monitored through mobile devices.
3. Integration Stage (2013–2018):
   With the increase in IoT devices, companies began integrating IoT into various industries, including smart homes, healthcare, and agriculture.
4. Maturity Stage (2019–Today):
   The maturity stage saw the integration of IoT into various aspects of our lives, from smart homes to smart cities and Industry 4.0.

Future of IoT:
From 2020 to 2027, the IoT market is anticipated to expand at a compound annual growth (CAGR) of 25.2%. The rising need for new IoT applications, such as linked automobiles and smart factories, is the reason for this rise.

The concept of IoT has been around for decades, and over time, it has evolved to become an integral part of our lives. From smart homes to smart cities and Industry 4.

2.4   IMPORTANCE AND CURRENT RELEVANCE

The IoT is a very hot topic right now. If you're not familiar with it, the concept refers to everyday devices being connected to the internet and controlled remotely.

IoT products are used in homes (such as smart thermostats, security cameras, and baby monitors), businesses, and factories. The IoT trend will grow as technology improves and more everyday objects become connected.

Some major players in IoT are Amazon, Google, and Apple. These companies are constantly developing new ways to connect everyday items over the Internet. For example:

Siri voice commands are used in Apple's HomeKit to control their smart home devices.

Amazon's Echo connects to various smart home devices, allowing users to control them using voice commands.

Google is currently developing its own smart home platform called Google Home.

2.5   IOT TECHNOLOGIES AND ARCHITECTURE

An ecosystem of equipment and gadgets that are linked to the Internet and can communicate with one another is referred to as the IoT.

Data from IoT devices is communicated, examined, and used to make decisions. Sensors, data storage, communication protocols, and applications make up the layers of an IoT architecture.

IoT architecture includes three layers:

Sensors and devices: These gather information from the real world and transmit it to the IoT platform.

Protocols for communication: Data is sent between devices and the cloud via these protocols. The most widely used protocols are Bluetooth Low Energy (BLE), Wi-Fi, and Zigbee.

Data processing and storage: The databases and analytical tools that make up this layer are used to store and handle the data that IoT devices have acquired. Additionally, it offers machine-generated insights and user interfaces for decision-making.

2.5.1   SENSORS AND ACTUATORS

Sensors and actuators are key components in various systems, particularly in the fields of engineering, automation, and control. They play an important role in collecting and processing data, as well as driving physical actions in response to that data.

Sensors are instruments that are capable of detecting and quantifying various physical events or attributes, including motion, light, humidity, pressure, and temperature. They transform the signals they detect into electrical signals or digital data that other system components can process with ease. It is usual to find temperature, pressure, motion, proximity, and ultrasonic sensors, among other sorts of sensors.

Actuators, on the other hand, are devices that convert electrical signals or digital commands into physical actions. They are responsible for controlling and moving physical systems, such as opening and closing valves, adjusting motor speeds, or operating robotic arms. Examples of actuators include motors, solenoids, relays, pumps, and pneumatic or hydraulic cylinders.

In many applications, sensors and actuators work together in a closed-loop system. Sensors detect and measure certain physical parameters, and the obtained data is processed by a controller or a computer. The controller then sends appropriate commands to actuators to perform specific actions based on the collected data. This feedback loop allows systems to continuously monitor and adapt to changes in their environment, ensuring efficient and accurate operation.

Numerous industries and applications, including robotics, aircraft, automotive systems, industrial automation, healthcare, and smart homes, use sensors and actuators. They make it possible to create intelligent, self-governing systems that can communicate with the outside world and react instantly to changes in their surroundings.

2.5.2 Connectivity and Network Protocols

The capacity of various networks and devices to connect and communicate with one another is referred to as connectivity. In order to facilitate data interchange and transmission, it entails establishing a steady and dependable link between devices or networks. Figure 2.1 shows the IoT protocols.

The sort of data being transmitted and the particular network requirements determine which protocols are used.

2.5.3 Edge and Cloud Computing

While cloud computing and edge computing are closely related, they have various uses in the computing industry.

The process of processing and evaluating data at or close to the location where it is generated, as opposed to sending it to a central data center or the cloud, is known as edge computing. Real-time data processing is made possible by this method, which lowers latency and speeds up reaction times. Edge computing is especially helpful in situations when network bandwidth is limited, including in remote or mobile contexts, or when low latency is essential, like in industrial IoT (IIoT) applications or autonomous vehicles.

On the other hand, cloud computing involves using a network of remote servers to store, manage, and process data rather than utilizing local servers or personal computers. This allows for scalable and on-demand access to computing resources, enabling organizations to easily provision and deploy applications and services

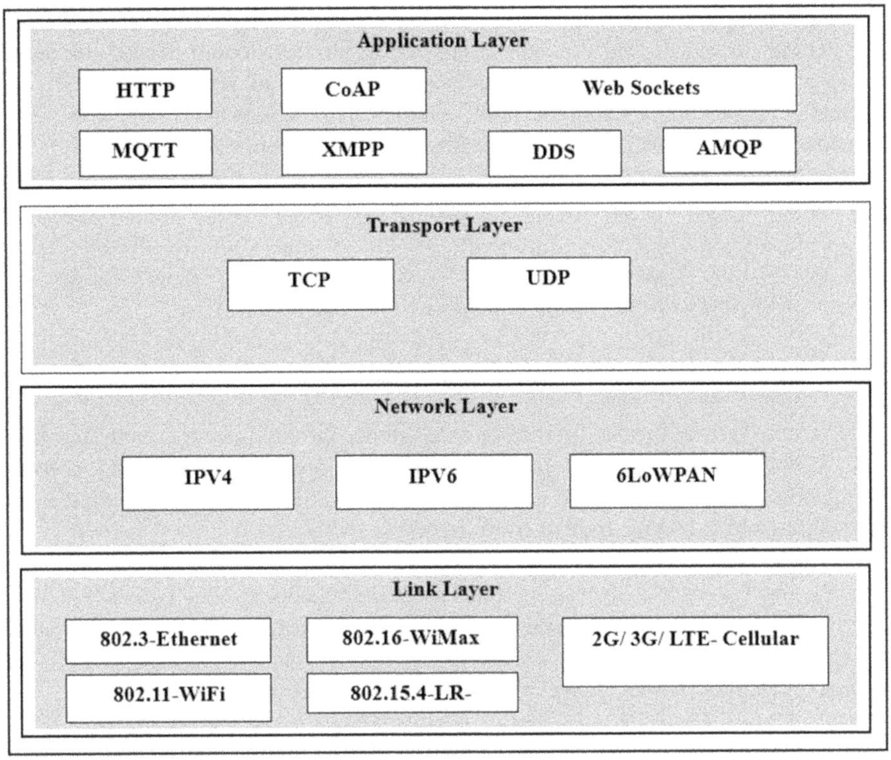

FIGURE 2.1   IOT protocols.

without having to invest heavily in infrastructure. Cloud computing provides high availability, resilience, and flexibility for various workloads and allows data to be accessed from anywhere with an Internet connection.

Edge computing and cloud computing are complementary to each other and can work in tandem to provide a comprehensive computing solution. Edge devices can collect and process data at the edge, while the cloud can provide storage, data analysis, and management capabilities. This combined approach is known as fog computing, where a hybrid architecture of edge and cloud resources is used to efficiently process and manage data.

Overall, edge computing focuses on decentralized data processing and real-time decision-making, while cloud computing provides scalable and centralized computing resources. Each strategy has benefits of its own and can be applied depending on the particular needs of a use case or application.

2.6   IOT PLATFORM AND MIDDLEWARE

The creation and implementation of IoT solutions require IoT platforms and middleware. They offer the infrastructure, services, and tools required to link and control the different systems and gadgets in an IoT network.

IoT platforms serve as a central hub for collecting, processing, and analyzing the data generated by IoT devices. They typically offer functionalities such as device management, data management, real-time analytics, and integration with other systems or applications. These platforms enable organizations to monitor and control their IoT devices, as well as derive meaningful insights from the collected data.

Middleware, on the other hand, acts as a bridge between the IoT devices and the applications or services that utilize the data. It provides services such as data transformation, protocol conversion, message routing, and security. Middleware effectively manages the communication and interaction between IoT devices and backend systems, allowing seamless integration and interoperability.

2.7   IOT COMPONENTS AND ECOSYSTEM

The IoT ecosystem consists of various interconnected components, including hardware devices, communication protocols, network protocols, APIs, and software frameworks. These devices and protocols enable IoT devices to communicate with one another and exchange data over the Internet.

Message Queuing Telemetry Transport (MQTT), Constrained Application Protocol (CoAP), and Extensible Messaging and Presence Protocol (XMPP) are a few of the widely used communication protocols in the IoT ecosystem. These protocols allow IoT devices to communicate with servers, gateways, and other IoT devices over various types of networks, including Wi-Fi, cellular, and satellite networks.

2.8   DEVICES AND HARDWARE

Some common devices and hardware in IoT include:

1. Sensors: Various factors, including temperature, humidity, light, pressure, motion, and more, can be measured and collected using these devices. Data from the physical world is captured in real-time with the use of sensors.
2. Actuators: Actuators are devices that enable physical actions or changes based on the data received from sensors. They convert electrical signals into mechanical movements, such as opening/closing valves, turning switches on/off, adjusting positions, etc.
3. Microcontrollers: These are small computer chips that integrate microprocessor, memory, and programmable input/output peripherals. Microcontrollers are often used to control and manage IoT devices, enabling them to process data and execute commands.
4. Communication modules: These modules facilitate wireless connectivity, allowing IoT devices to communicate with each other and with the central server or cloud-based platforms. Common communication protocols used in IoT include Wi-Fi, Bluetooth, Zigbee, cellular networks (2G, 3G, 4G, and 5G), and LoRaWAN.
5. Embedded systems: Embedded systems are hardware and software components integrated into a larger IoT device or system. They are designed for specific

functions and are often low-power, resource-constrained, and specialized for a particular application.

6. Cloud-based platforms: While not physical devices, cloud-based platforms are an essential component of IoT infrastructure. They provide storage, computing power, and analytics capabilities, enabling the processing and analyzing vast amounts of IoT data.

7. Robotics and automation systems: IoT-enabled robots and automated systems use sensors, actuators, and connectivity to perform tasks autonomously or with minimal human intervention. These devices find applications in industries like manufacturing, healthcare, agriculture, logistics, etc.

8. Smart home devices: Smart appliances, door locks, security cameras, lighting controls, and thermostats are examples of IoT devices that are part of the smart home. They may be operated remotely with voice assistants or smartphones by connecting them to a home network.

Overall, the devices and hardware used in IoT systems enable the collection, transmission, processing, and control of data, making IoT applications possible across various domains.

2.9   DATA ANALYTICS AND AI IN IOT

1. Anomaly Detection: Data analytics tools and AI algorithms help in detecting anomalies in the IoT data. Unusual patterns or deviations from normal behavior can indicate potential issues, security breaches, or anomalies in the system. Early detection of anomalies can prevent potential failures or threats to the IoT infrastructure.

2. Energy Optimization: With the help of data analytics and AI, IoT devices can optimize energy consumption. By analyzing data on energy usage patterns, organizations can identify inefficiencies and implement measures to optimize energy consumption, reduce costs, and improve sustainability.

3. Personalization and User Experience: By analyzing user behavior and preferences, IoT devices combined with AI offer personalized experiences. With data analytics, organizations can derive insights about user preferences, habits, and choices, enabling them to deliver personalized services and enhance the overall user experience.

4. Cybersecurity: IoT devices are prone to cybersecurity threats due to their interconnected nature. AI and data analytics can help identify and mitigate potential security breaches or anomalies. Machine learning algorithms can detect abnormal patterns in data traffic and alert system administrators.

Data analytics and AI are critical components in leveraging the full potential of IoT. By collecting, analyzing, and interpreting the enormous amount of data generated by IoT devices, organizations can optimize operations, improve decision-making, enhance user experience, and ensure the security of their IoT ecosystem.

2.10 SECURITY AND PRIVACY CONSIDERATIONS

You should also consider buying a security camera that encrypts your video feed while it's being transmitted. That's especially important if you have neighbors or others who can intercept your Wi-Fi signal.

Additionally, you should only share your camera's video feed with people you trust. If you have an outdoor camera pointed at your front door, don't share that link with strangers via social networks or messaging apps.

If you want more security and privacy, consider buying a security camera that stores its footage locally on an SD card or a hard drive rather than in the cloud. That's the only way to guarantee that your footage won't be accessed by hackers or other people without your permission.

2.11 IOT IN DAILY LIFE

With its incredible characteristics, the IoT is changing the world. It is a network of actual physical objects, including buildings, cars, and other things, that are linked to the Internet and can gather and exchange data. IoT devices are capable of communicating, sharing data, and carrying out activities without the need for human assistance.

IoT is applied in several industries, including home automation, transportation, healthcare, and agriculture. Here are a few instances of IoT applications in daily life:

1. Smart houses: To improve comfort, efficiency, and energy conservation, smart houses employ IoT devices. For instance, smart lighting may automatically adapt based on occupancy and preferences, and smart thermostats can be controlled remotely.
2. Health Monitoring: Biometric data, such as heart rate, blood pressure, and sleep habits, are tracked by IoT-enabled health monitoring devices.
3. Vehicle Telematics: IoT devices are utilized in automobiles to offer functions, such as emergency assistance, vehicle tracking, and remote locking and unlocking.
4. Smart Agriculture: IoT devices are used to monitor cattle, crops, and soil conditions.

2.11.1 SMART HOMES AND CONSUMER APPLICATIONS

A smart home is an automated system that uses Internet-based technology to manage the lights, appliances, temperature, and other elements of a house. The idea behind a smart home is to combine several gadgets into a single, controllable system that can be accessed through a mobile app or central hub.

The concept of a smart home is relatively new, and many homeowners are excited about its potential to enhance their lifestyle. However, the concept of a smart home is not limited to just residential use.

A smart home can also be applied to the commercial sector. For example, businesses can use smart technology to control various aspects of their operations, such as energy usage, security, and employee productivity.

2.11.2 WEARABLES AND PERSONAL IoT DEVICES

Wearables and personal IoT devices have been making headlines in recent years and are becoming increasingly popular. These devices are Internet-enabled and can connect to the Internet using WiFi or cellular networks.

Smartwatches, fitness trackers, and augmented reality glasses are examples of wearable technology that is intended to be worn on the body to gather information about an individual's activities and bodily processes.

Devices classified as personal IoT are those that are usually not wearable and intended for usage in certain settings or contexts.

These gadgets can be personal medical equipment like blood pressure cuffs, glucose monitors, and heart rate monitors, as well as smart home gadgets like security cameras, smart lighting, and smart thermostats.

Wearables and personal IoT devices have a lot of advantages. Users who use these gadgets can increase productivity, maintain relationships with friends and family, and lead healthier lives.

2.11.3 SMART CITIES AND URBAN DEVELOPMENT

There are several key areas in which smart cities focus on urban development:

1. Infrastructure: Smart cities invest in advanced infrastructure to support sustainable and efficient operations. This includes using smart grids for electricity distribution, intelligent transportation systems for traffic management, and integrated sensor networks for monitoring and managing utilities such as water supply and waste management.
2. Mobility: Smart cities aim to offer sustainable and effective transit choices. To lessen traffic, improve air quality, and increase accessibility for all citizens, they use technology, including shared mobility services, intelligent traffic management systems, and real-time public transportation tracking.
3. Public safety: Modern technologies are used in smart cities to improve security and safety for the general public. This includes employing real-time emergency response systems for quicker assistance during emergencies, predictive analytics to pinpoint crime hotspots, and facial recognition technology in video surveillance systems.
4. Environment: Smart cities prioritize sustainability and environmental conservation. They deploy smart energy grids, implement renewable energy sources, and promote energy-efficient buildings. They also use data analytics to monitor air quality, reduce waste generation, and improve water management.
5. E-Governance: Smart cities leverage technology to enhance citizen engagement and improve the delivery of government services. They develop online platforms and mobile apps for citizens to access information, make payments, and provide feedback. These cities also use data analytics to identify patterns and make data-driven policy decisions.

Urban development and smart cities work hand in hand to build livable, sustainable, and resilient urban environments through efficient use of data and technology. Smart cities improve the quality of life for their citizens by integrating e-governance techniques, public safety measures, mobility solutions, and modern infrastructure.

2.11.4 INTERNET OF THINGS IN INDUSTRY

The sector could undergo a change thanks to the IoT, which could make asset tracking, process monitoring, and operational optimization easier and more affordable for businesses.

2.11.5 INDUSTRIAL IoT (IIoT) AND MANUFACTURING

Businesses are using IoT technologies to increase productivity and efficiency as they continue to automate and modernize their processes.

IIoT and manufacturing are two areas where IoT is having a big influence on industry.

IIoT refers to the process of improving operations and boosting production by connecting industrial equipment, such as machines, sensors, and control systems, to the Internet. Businesses may have real-time visibility into their operations by gathering and evaluating data from these devices. This allows them to optimize their procedures and make data-driven decisions.

IoT technologies are especially well-suited to the manufacturing sector, which uses a variety of sophisticated machinery and processes that may be made more efficient by gathering and analyzing data in real time. Manufacturers, for instance, can employ IoT-enabled devices.

2.11.6 AGRICULTURE AND ENVIRONMENTAL MONITORING

IoT in agriculture and environmental monitoring allows for the remote monitoring and analysis of crops, soil, and environmental conditions. IoT devices such as sensors placed on agricultural equipment, IoT devices in soil, and IoT devices in the atmosphere collect real-time data that farmers, researchers, and policymakers can use to make informed decisions.

IoT devices in agriculture and environmental monitoring can also be used to monitor air and water quality, which is essential for maintaining soil and water quality and protecting the environment. These devices can be used to identify and measure levels of pollutants, such as nitrogen and phosphorus, in water and soil, which can contribute to algal blooms and eutrophication of water bodies.

Overall, IoT devices in agriculture and environmental monitoring can help farmers, researchers, and policymakers make informed decisions that improve crop yields, protect the environment, and ensure the quality and safety of food and water resources.

2.11.7 Healthcare and Remote Monitoring

Among the sectors that stand to gain the most from IoT technology is healthcare. IoT can assist healthcare professionals in lowering expenses, enhancing patient care, and optimizing the use of scarce resources.

The ability to remotely monitor patients is one way that the IoT can benefit healthcare providers. IoT-enabled gadgets, for instance, can be used to monitor vital signs like blood pressure and pulse rate and transmit the collected data to medical professionals for review. This can assist medical professionals in detecting health issues early on, which can minimize the need for in-person visits and save expenses.

2.12 CHALLENGES AND FUTURE DIRECTIONS

By optimizing the flow of goods and services, businesses can provide their customers with faster delivery and better service.

2.12.1 Interoperability and Integration

In the IoT, interoperability and integration are two key ideas that are vital to facilitating effective data interchange and communication among diverse systems, applications, and devices. In this chapter, we will examine the definitions of interoperability and integration in the context of the IoT, as well as some possible advantages.

In the IoT, interoperability is the capacity of various systems and devices to exchange data and communicate with one another.

The concept of interoperability is based on the fact that different IoT devices, regardless of their manufacturers, should be able to talk to each other and share data seamlessly. In other words, interoperability enables different devices, applications, and systems within the IoT ecosystem to work together without requiring modifications or additional efforts.

In contrast, integration in the context of IoT refers to the process of integrating various systems and devices into a broader IoT ecosystem. Integration involves bringing together various IoT elements, such as sensors, devices, applications, and platforms, to create a cohesive and interconnected system. The goal of integration is to enable seamless data exchange and interaction among these components.

By implementing interoperability and integration in IoT, various benefits can be achieved:

1. Enhanced efficiency: Interoperability and integration in IoT allow different devices and systems to communicate with each other seamlessly, resulting in enhanced efficiency and reduced manual intervention.
2. Data integration: By integrating different devices and systems, IoT enables seamless data exchange between different entities within the IoT ecosystem. This facilitates data-driven decision-making, analytics, and actionable insights.

3. Scalability and interoperability: Interoperability ensures that IoT devices can integrate with other systems and devices, regardless of their manufacturers or communication protocols. This promotes scalability and adaptability, allowing the IoT ecosystem to grow and evolve without being limited by specific vendor lock-ins.
4. Reduced costs: Integration of IoT devices and systems can save costs by eliminating the need for separate systems or interfaces, streamlining operations, and optimizing resource allocation.

For devices and systems inside the IoT ecosystem to communicate and exchange data seamlessly, interoperability and integration are essential. By facilitating efficient communication, integration promotes scalability, cost reduction, and improved decision-making.

2.12.2 Scalability and Infrastructure

Infrastructure and scalability are essential components of the IoT. Scalability is the capacity to manage growing numbers of devices and data without appreciable performance reduction. Infrastructure is the supporting framework that enables connectivity between devices and the cloud.

IoT systems need to be scalable in order to expand and change to meet shifting needs. The system ought to be able to manage the rise in data and processing demands as more devices are connected to the network without experiencing overload.

Infrastructure, on the other hand, refers to the physical and digital infrastructure that supports IoT. This includes network connectivity, data centers, cloud infrastructure, and devices themselves.

2.13 CONCLUSION

We have discussed the key points, such as the ability to collect vast amounts of data, the seamless connectivity of devices, and the automation of mundane tasks, all of which contribute to the profound potential of IoT. This technology revolution has the potential to improve people's quality of life in addition to increasing operational efficiency for corporations. IoT has enormous potential effects on both the business and society.

Smart cities could lead to more sustainable and efficient living environments, while industries can achieve unmatched levels of productivity and innovation. From a societal perspective, the enhanced data and connectivity across devices can lead to improved healthcare, smarter transportation systems, and more personalized consumer experiences.

Looking forward to the future of IoT, we are on the cusp of a new era where every object can be smart and connected. The possibilities are limitless, and the opportunities for improvement and growth are boundless. While challenges such as security and privacy remain, with the right regulations and ethical considerations in place, IoT can lead humanity into a more interconnected and intelligent world.

As we stand at the threshold of this exciting future, the importance of embracing IoT cannot be overstated. It's time for policymakers, business leaders, and individuals alike to harness the power of IoT to shape a better tomorrow. So, let us take the first step toward unlocking the full potential of the IoT by educating ourselves, advocating for responsible use, and investing in this groundbreaking technology. Join the movement and be a part of the IoT revolution!

3 Unlocking the Power of IoT
An In-Depth Exploration

*Prakash Shanmurthy, Damodharan D.,
Anandhan Karunanithi, Pajany M., and
J. John Bennet*

3.1 INTRODUCTION TO IOT

An Internet of Things (IoT) network consists of devices and entities that are interconnected and can exchange information over the Internet. Anything from mundane objects like domestic appliances, automobiles, wearable technologies, and even industrial machinery to basic electrical devices like smartphones and laptops might be included in these gadgets. The primitive objective of IoT is to make it feasible for these entities to connect, communicate, and collaborate in a simple manner. This will improve convenience, automation, and productivity in many facets of daily living [1].

The revolutionary concept that alludes to a very great extent network of interrelated devices and items, all equipped with sensors, processors, and communication capabilities are called IoT. These devices may be used to gather and distribute information online without direct contact or a link between a person and a computer. Essentially, IoT is about making these gadgets connect, communicate, and work together seamlessly, increasing productivity, efficiency, and convenience in many aspects of our lives [2].

In IoT, "Things" can gamut from traditional electronic tools identical to smartphones, computers, and suitable for wearing, to everyday objects like home appliances, vehicles, industrial machines, and even sensors embedded in the environment. In the IoT, there are a number of connected devices that enable the production, analysis, and utilization of real-time data.

In recent years, miniaturization, sensor technologies, cloud computing, and data analytics have contributed to the growth of the IoT. With these elements, IoT can be integrated into a various fields, including smart cities, smart homes, healthcare, transportation, agriculture, and manufacturing [3].

The IoT is changing the way humans interact with the world by transforming inanimate objects into "smart" devices that are able to communicate, sense, and respond to their environment and human interactions. An increasingly connected and data-driven world will hold new opportunities for innovation and efficiency as the IoT continues to evolve [4,5].

DOI: 10.1201/9781032623276-3

3.2 DEFINITION AND EVOLUTION OF IOT

The concept of connecting devices and enabling them to communicate has existed for much longer than the term "Internet of Things" was coined. Several phases can be identified in the evolution of IoT:

- Early Concepts and Radio Frequency Identification (RFID) (1990s–2000s): As early as the late 1990s and early 2000s, the idea of connecting physical objects to the Internet was explored. The initial focus was on RFID technology, which made exploitation of radio signals to track and identify items. However, the breadth and realization of these early conceptions were still incomplete [3].

- Proliferation of Internet and Connectivity (2000s): As Internet connectivity became more widespread and affordable, the groundwork for IoT was laid. In recent years, high-speed Internet and wireless communications, such as Bluetooth and Wi-Fi, have significantly improved connectivity and communication between devices [6].

- Emergence of Smart Devices (2010s): Smart devices with sensors, processors, and networking capabilities gave rise to IoT in the 2010s. Smartphones, smartwatches, activity trackers, and smart home appliances were among these gadgets. The IoT ecosystem is a result of these devices' access to the Internet and to one another.

- Machine-to-Machine (M2M) Communication (2010s): IoT has become a major factor in the development of M2M communication, a method of transferring data directly between devices without human involvement. This enabled automation and data exchange between devices, leading to improved efficiency in various industries, such as manufacturing and logistics [1,3].

- Big data and Cloud computing (2010s): Big data and cloud computing technologies were necessary to handle the enormous amounts of data generated by IoT devices. An affordable and scalable cloud service provides scalability and storage. With big data analytics, it is now possible to gain valuable insights from this data and to enable data-driven decisions to be made based on that information [6,4].

- Intelligent IoT and Artificial Intelligence (2010s): The consolidation of artificial intelligence and machine learning with IoT devices brought about the concept of intelligent IoT. Smart devices became capable of analyzing and interpreting data on their own, making autonomous decisions, and adapting to user preferences. This paved the way for applications like smart assistants, predictive maintenance, and personalized user experiences [7,8].

- Industrial IoT (IIoT) and Smart Cities (2010s): IoT applications grew beyond consumer gadgets as the technology developed. Manufacturing, energy, agriculture, and healthcare are just a few of the areas that have been transformed by the IIoT idea. Similar to this, smart cities have gained a lot of attention. They use IoT technology to improve urban infrastructure, streamline municipal services, and raise inhabitants' quality of life in general [8].

- 5G Connectivity (2020s): The rollout of 5G technology in the 2020s has further accelerated the evolution of IoT. With 5G, real-time communication is enhanced, connected devices are enabled, data transfer speeds are faster, with low latency, and connections are more plentiful [7,8,3].

- As technology continues to advance and more devices become connected, a more connected and data-driven world is anticipated as a result of the IoT ecosystem's predicted exponential growth and new breakthroughs and uses across several industries.

3.3 KEY COMPONENTS OF IOT

The IoT comprises various key aspects that operate collectively to facilitate seamless connectivity, data exchange, and interaction between devices. These components form the foundation of the IoT ecosystem. The essential factors of IoT include:

1. Device/Things: Devices, also known as "Things" in IoT, are physical objects that are equipped with sensors, actuators, and communication modules. These devices can be anything from intelligent phones, sensing devices, wearable technologies, and domestic appliances to modern machinery, vehicles, and environmental monitoring equipment.
2. Connectivity: The significant IoT-based system assists in the connectivity to communicate and interchange data with each other. Various communication technologies facilitate this, incorporating Bluetooth, cellular networks (3G, 4G, 5G), Wi-Fi, Z-wave, LoRaWAN, Zigbee, and NB-IoT [6,4,3].
3. Data Processing: A huge amount of data is produced by IoT from linked devices. Data processing is gathering, storing, and analyzing this data in order to get an insightful understanding and make defensible judgments. Data processing might take place in centralized cloud servers or on the device itself (edge computing).
4. Cloud Computing: It provides scalable and secure storage and processing capabilities for IoT data. Cloud platforms enable data aggregation, analysis, and storage, making it accessible from anywhere and facilitating real-time data exchange.
5. Data Analytics and Artificial Intelligence (AI): IoT devices spawned substantial volumes of data in a critical task for Statistical analytics and AI [3]. Automated decision-making, personalized suggestions, and predictive maintenance are made possible by AI algorithms' ability to analyze data patterns, spot abnormalities, and gain insightful knowledge [7,8,4].
6. User Interface: Interacting with IoT systems and devices is made possible by the user interface component. This can be done using voice assistants, web interfaces, mobile applications, or other graphical user interfaces, making it simpler for consumers to manage and keep an eye on linked devices [6].
7. Security and Privacy: It is critical that IoT devices are secure, as they are susceptible to cyber threats and data breaches. Scrupulous protective assessment, including encryption, endorsement, and acquire controls, are necessary toward guarding IoT devices and the information they collect [9,2,8].
8. Standards and Protocols: IoT devices need to adhere to standard communication protocols to ensure interoperability and seamless data exchange.

Standardization efforts, such as Message Queuing Telemetry Transport (MQTT), Constrained Application Protocol (CoAP), Hypertext Transfer Protocol (HTTP), and Open Platform Communications Unified Architecture (OPC-UA), help create a unified IoT ecosystem [3].

9. Edge Computing: In its place of transmitting all data to federal cloud servers, edge computing administers data farther from where it originates (at the edge of the network). Edge computing reduces latency, bandwidth usage, and optimizes real-time adaptability, rendering it perfect for IoT applications that require quick response times [7,8].

10. Firmware and Over-the-Air (OTA) Renews: Firmware upgrades for IoT devices are frequently needed to incorporate new functionalities, fix bugs, or enhance security. OTA updates allow devices to be updated remotely without physically accessing them.

These essential IoT elements work together to provide a connected ecosystem that enables devices to interact, collect and analyze data, and offer intelligent and automated services. As a result, efficiency, convenience, and innovation are ultimately increased across a variety of sectors and disciplines [5].

3.4 APPLICATION AND USE CASE OF IOT

There have been a number of applications of IoT in a variety of domains, including the following:

1. Smart Homes: Home automation is made possible by IoT, allowing users to operate appliances like smart thermostats, lighting controls, and security cameras remotely using their smartphones or voice commands.

2. Healthcare: IoT devices such as wearables and remote monitoring systems help in tracking health parameters and providing real-time health data to healthcare professionals, leading to better diagnosis and personalized treatment [7,6].

3. IIoT: IoT is utilized in manufacturing and industry to enhance efficiency and decrease downtime through proactive maintenance, process optimization, and asset tracking [8,5].

4. Smart cities: The IoT is now being used in a variety of ways in order to improve urban infrastructure, control traffic, optimize energy consumption, and advance overall city services [2,7].

5. Agriculture: IoT-enabled sensors in agriculture support optimizing irrigation and increasing yields, keeping an eye on crop fitness, climate patterns, and soil specifications [7].

3.5 CHALLENGES AND OPPORTUNITIES IN IOT

3.5.1 CHALLENGES

Huge prospects for modernization and connection are provided by IoT. Nonetheless, it also poses a number of difficulties that must be resolved to guarantee its implementation's success and security. IoT is currently facing a number of challenges, some of which are listed below:

- Privacy and Security: As IoT devices often transmit and gather sensitive information, they are vulnerable to hacking and privacy violations. Unauthorized data retrieval or manipulation of connected devices might result from inadequate security measures and inadequate encryption [9,2].
- Interoperability: The IoT ecosystem consists of numerous devices and platforms from different manufacturers. The development of cohesive IoT solutions can be hindered by the difficulty of ensuring seamless communication and compatibility between these devices.
- Scalability: IoT network's infrastructure management and scalability become a rise in difficulty as additional linked devices are added. As the number of devices increases, it is more difficult to ensure that the system remains efficient and reliable.
- Data Management and Analytics: IoT creates a tons of data from many sources. It might be difficult to adequately organize and interpret this data in order to produce practical insights; this requires strong data analytics capabilities [7,2,4,3].
- Power Consumption: The majority of IoT devices are battery-powered, and optimizing power consumption is essential in order to extend their lifespan and reduce the frequency with which they need to be replaced or recharged.
- Connectivity Issues: Internet access must be dependable and consistent for IoT. Data transfer and device control may be difficult in places with patchy network coverage or inconsistent connectivity.
- Standards and Regulations: The lack of universal standards and regulations in the IoT industry can create challenges in terms of data security, interoperability, and compliance with legal requirements [3].
- Cost: IoT infrastructure and device deployment can be expensive at first, especially for large-scale applications. Widespread adoption may be hampered by this expense, particularly in certain sectors or geographical areas [4].
- Privacy Concerns: IoT devices are embedded in various aspects of daily life, raising concerns about constant data collection and potential invasion of privacy [9,2,3,10].
- Environmental Impact: When IoT devices proliferate quickly and are not properly recycled or disposed of this can result in electronic waste and environmental problems [7].
- Reliability and Resilience: IoT devices and systems need to be highly reliable and resilient, notably in mission-critical settings like medical care, transportation, and business operations where failure can have serious repercussions [3].

Collaboration among a variety of stakeholders, including producers, legislators, researchers, and consumers, is necessary to address these issues. Strong security measures, adherence to privacy laws, and the development of open standards can all help to address some of the major IoT concerns and unlock the full potential of the technology across a variety of industries [7].

3.5.2 OPPORTUNITIES

An array of opportunities are offered by the IoT across a wide range of industries and fields. As technology develops and linked gadgets keep multiplying the opportunities in IoT become increasingly compelling. Some of the key opportunities in IoT include:

Enhanced Efficiency: The IoT enables automation and real-time data analysis and improves the efficiency of processes and operations. Some of the efficiency-driven opportunities in IoT include optimizing operations, eliminating waste, and optimizing resource consumption.

Improved Decision Making: Through its network of linked devices, IoT produces colossal volumes of data. Businesses and groups may be able to maximize their strategies and make decisions based on data by examining this data and gaining insightful insights.

Personalized and Contextual Experiences: By gathering and analyzing information provided by connected gadgets, IoT can enable customized services for consumers. This makes it possible to provide customized services like custom suggestions, automated smart homes, and targeted advertising [2].

Healthcare Advancements: Remote tracking of patients, wearable medical technology, and telemedicine services are all ways that IoT is revolutionizing the healthcare sector. It enables proactive and individualized treatment, improving patient outcomes, and lowering healthcare expenditures [7,6].

Smart Cities: IoT is revolutionizing urban living by making cities smarter, more efficient, and sustainable. Smart cities cover a wide range of topics, including smart mobility, energy management, waste management, and public safety [2,7].

Industry 4.0 and Smart Manufacturing: IoT is driving the fourth industrial revolution, where connected machines, sensors, and data analytics enable smart manufacturing and intelligent supply chain management. As a result, production rises, downtime decreases, and the quality of the final product improves [8,5].

Agricultural Innovation: IoT-based precision agriculture is transforming farming practices, enabling farmers to monitor crops, optimize irrigation, and improve resource utilization. This promotes secure access to food and sustainable agriculture.

Connectivity and Communication: IoT makes it possible for people and things to communicate and interact seamlessly. This enables real-time collaboration, remote monitoring, and improved communication channels.

Environmental Monitoring and Conservation: IoT devices are capable of detecting animals and monitoring environmental factors, such as the quality of the air and water. Using this knowledge, the ecosystem might be conserved, and the impacts of climate change may be mitigated [7].

Safety and Security: IoT technologies provide improved security and safety measures, including surveillance, asset monitoring, and smart home security systems. These advancements help in preventing accidents, reducing crime rates, and protecting valuable assets [2].

Smart Energy Management: Energy use in homes, buildings, and enterprises may be better controlled and monitored owing to the IoT. This leads to optimized energy usage, cost savings, and reduced carbon emissions [6].

Innovation and Entrepreneurship: The growth of IoT presents numerous opportunities for entrepreneurs and startups to develop innovative solutions and disrupt existing industries.

IoT delivers a wide range of options for organizations, individuals, and communities. Embracing IoT technologies and leveraging their potential can lead to transformative changes in how we live, work, and interact with the world around us.

3.6 ARCHITECTURAL LAYERS OF IOT

3.6.1 INTRODUCTION

IoT architecture is the combination of sensors, software, communication protocols, vehicles, intelligent devices, buildings, physical objects, cloud services, users, developers, business layers, actuators, etc. Data management systems gather data for storage, analysis, processing, and exchange. All these network physical components, configuration, practical organization measures, and working values.

3.6.2 REAL-WORLD COMPONENTS OF IoT

The practical components of IoT encompasses of several layers. They are:

1. Application and service
2. Detection
3. Announcement
4. Cloud
5. Administration

3.6.2.1 Applications and Service

This layer provides a variety of services, which include the collection of data, analyzing the data, visualizing the data, and providing security to the data. All these depend on the use cases and features of the end users.

3.6.2.2 Detection

An important function of the sensing layer is to detect fluctuations in the physical form of the practical components during actual stretching. The main component of this layer is sensors, which are responsible for recognizing and positioning the intellectual objects, gathering the information, and transporting the information to the cloud environment for processing and storing the information. In this layer, we can find an actuator, which is a mechanical device such as a switch that is used to execute the chosen device.

3.6.2.3 Announcement

The communication layer has certain devices like gateways, routers, and switches, which are connected to devices that cannot link to the cloud services directly. This layer is accountable for interactions or transfer of information to the other IoT layers. The data which are composed by the sensing layer will be shifted to the cloud service and the application layers.

3.6.2.4 Cloud

Universally, the function of the cloud is to use a data center as a fundamental server to develop the generated information by the edge device. Information composed from sensors and devices are consumed into the cloud layer. The purpose of cloud is to store data, processing of data, and finally analyze the information that is available. In different circumstances we can say this cloud layer as the IoT system processing unit.

3.6.2.5 Administration

The responsibility of the administration layer is to operate and monitor all the other layers with the help of cloud administration tools, which are generally employed [11].

3.6.3 IoT Architecture

Designing a flexible architectural outline is vital for IoT module at networking with huge amount of diversity over the Internet. The major concerns to be considered while developing the architecture of an IoT network are

- Privacy and security
- Data storage and Quality of Service (QoS)
- Scalability
- Reliability

The absence of a solid architectural pattern has led to numerous initiatives. However, no standard model has been adopted due to the wide range of applications and outline plans with diverse variables and design patterns. In order to scale the computation of the current classifications, the standard model has not been implemented.

Despite this, the most commonly used IoT architectures are:

1. Three-layer Architecture.
2. Middleware-based Architecture.
3. Service-Oriented Architecture.

3.6.4 THREE-LAYER ARCHITECTURE

IoT research began with the development of the three-layer architecture, which has been referred to as the most fundamental architecture. The architecture is divided into three layers, such as

1. The application layer describes how the computer software contributes to the exact provision of a service.
2. In the network layer, information is broadcast to the upper layers.
3. The Perception Layer describes the physical surface of an object and clusters all its features together.

3.6.4.1 Application Layer

The IoT Architecture begins with the application surface. It lies in between the user program and the application program. A network layer administrator is responsible for retrieving information from a network layer and utilizing information that has been composed for explicit purposes. The information will be used for warehousing, accumulating, filtering and managing, databases, etc. As an outcome of this, the information will be accessible to applications of IoT such as intelligent cars, wearables, intelligent cities, intelligent transportation, intelligent agriculture, healthcare systems, etc. Furthermore, this layer analyzes the data that has already been received in the application layer and stored there to forecast the future condition of physical objects [6]. The most common software technologies that widely use this massive amount of information by the appliances are edge and cloud computing. The arrangement of the information to be processed can be any type as below:

- Small and binary information cannot be read by human beings;
- Larger text information is readable by human beings [12].

3.6.4.2 Network Layer

This network layer may be referred to as a communication layer, which is merged with the middle layer of the IoT, which is considered to be the fundamental component of the IoT ecosystem. Providing routing frequencies for information transmission should be the responsibility of the communication layer. The network layer is responsible for providing connectivity, transferring messages between various devices, cloud service providers, and countless servers. This layer acts as a boundary for perception and application layers, which helps IoT devices, applications, and servers to receive the information. Network layer consists of different communication standards, such as Bluetooth, Wi-Fi, and IoT objects. These devices will be useful in critical circumstances with the help of wireless protocols, which play an important role in this layer because they require less material and human effort. In addition to routing capabilities and unique addressing, IoT promises easy integration of various devices into the same device [13].

3.6.4.3   Perception Layer

The persistence of this layer is achieved by collecting certain information about the objects, and it has to identify the objects that are unique in the IoT ecosystems. This layer consists of sensors that transmit and receive information from other layers of the atmosphere, as well as processing the information in the upper layers [14]. The perception layer embraces objects that can communicate with peripheral devices which can sense with some computing capabilities, in general terms, we can say that as smart technologies.

IoT uses these smart technologies as the fundamental blocks that can be used for common purposes (smart watches, smart TVs, etc.). Smart objects were equipped with some general properties like sensing and actuation, addressability, communication, embedded data processing, identification, localization, and user interface. All these properties use a common factor called sensors to collect and communicate the information with other layers. Examples: Arduino, Raspberry Pi, Beaglebone Black, etc.

3.6.5   SECURITY AND CONFIDENTIALITY

Every system will have confidentiality and security issues that should be addressed to protect the IoT applications. There are some security issues associated with each layer of the IoT architecture, which should be taken into consideration at the very beginning of the design process. We will discuss some security issues in the IoT architecture at each layer.

3.6.6   SECURITY ISSUES IN THE APPLICATION LAYER

As per the necessities of the applications, this layer requires different security standards. Some listed privacy and security issues of this layer are:

1. Application-specific liabilities: Some liabilities may be left which was unknown to the users, which also creates a security threat to the user's information.
2. Information administration: There is an increase in complexity of the system as the collection of data increases, leading to a demand for creating resources and requiring complex algorithms to administer the information, which may lead to information loss.
3. User information confidentiality: At each communication, users' private information should be protected from vulnerable attacks. Sometimes, the technique which is used to process the information may lead to vulnerable activities, which may lead to huge loss of information.
4. Mutual confirmation and node identification: There are various levels of access privileges required by each application due to the presence of various users. It is essential to apply proper authentication procedures in order to prevent unauthorized access.

3.7 COMMUNICATION PROTOCOL FOR IOT

An important technology of the IoT is ZigBee, which enables users to benefit from smart applications and services. However, IPv6 transports an extreme of 1280 bytes of packets, ZigBee empowered instruments cannot have a grip on it. By using a gateway, these IPv6 packets will be transmitted over ZigBee networks and may pass through ZigBee Coordinator [15]. The main function of this ZigBee Coordinator and gateway is to discover the process which are close to the network. This creates a complexity problem for the ZigBee Coordinator. In addition to these routing structures, data forwarding issues, and header size problems are also there for ZigBee devices, which needs to get addressed [3,10–19]. As a solution to these issues for all IoT-enabled ZigBee devices, we can propose a peer-to-peer communication protocol hosted using the IPv6 over Low-Power Wireless Personal Area Networks (6LoWPAN) protocol over the Internet, which reduces the burdens on the ZigBee Coordinator.

The gateway used here will act as a protocol translator for passing the messages through with hosted Internet and with ZigBee node. It decodes the received messages to ZigBee frames, which originate as IPv4 packets travel through IPv6 packets, which demands hardware and software support, including the ZigBee devices that suffer from a header size problem [17].

Access to the whole network will be with the coordinator node. Once the coordinator node fails to function, it needs to be re-booted, which is not likely to execute, and it leads to the failure of the whole network in a single point of failure [15].

3.7.1 THOUGHT-PROVOKING PROBLEMS WITH ZIGBEE-ENABLED DEVICES

1. The peer-to-peer transmission with Internet hosts and ZigBee nodes.
2. A coordinator node controls the Zigbee network.
3. Hardware and software need fresh entry point.
4. Header dimensions problem.
5. Single point of failure.

To overcome these issues in an IoT-based network, we can use the 6LoWPAN communication protocol, which attaches a group of Zigbee devices with an Internet protocol-based setup that affords peer-peer announcement to direct ZigBee-enabled IoT networks with the IPv6 [18].

Three primary services will be offered when the protocol has an enquiry from the user, they are:

i. Disintegration and refabrication used.
ii. Title firmness.
iii. Link-layer forwarding.

3.7.2 ZIGBEE COMMUNICATION PROTOCOL

ZigBee protocol enables transmission of information with the edge nodes and the claim outlines, which supports low power protocol and low cost. This protocol

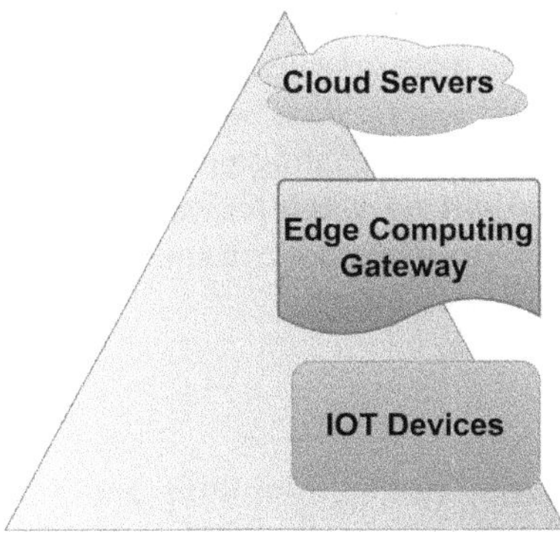

FIGURE 3.1   ZigBee protocol layer.

consists of five different layers that perform different functionalities to provide the proper communication between the nodes which are all connected [19]. The ZigBee Protocol layer is shown in Figure 3.1.

The five building blocks used to develop the ZigBee communication protocol are:

1. Application or Request Block
2. Network or Mesh Block
3. 6LoWPAN Block
4. Media Access Control (MAC) Block
5. Physical Layer (PHY) Block

3.7.3   6LoWPAN Block

The block is also called the 6LoWPAN procedure block, which is an amendment block and provides amenities to the Internet blocks used to collect the information which are available at the end nodes via MAC and PHY block, consolidates the collected information and handovers it over to the next upper layer. ND is used to find the neighbor nodes after that create the default communicating routes with the nodes and then the transmission starts [15].

The services provided by this protocol are shrinking header files, disintegration and refabrication, programmed formation, and neighbor detection in the upper block. Using shared information, the shrinking of header files service reduces the size of User Datagram Protocol and IPv6 header fields, which can be misplaced when header fields are removed from associated blocks [19].

3.7.4   MAC Block

The block offers three mechanisms: (i) Guided mode, (ii) Unguided mode, and (iii) General Switching Transceiver (GST) distribution. To accomplish non-breakable communication, communicating devices depend on the salutation, which can be done without signals. It will be helpful to save a huge amount of power as emitting signals at each and every triggering of the device will loss huge amount of power [18]. ZigBee communication networks use guided transmission frequency access to expand their potential and extend their snooze intervals. The routers in ZigBee occasionally release signals at all signaling points to broadcast their presence. The communicated signal edges will work as a register, which varies between 15 Ms and 252 s. Among each signal break, the nodes can snooze, dropping duty routines, which increases battery life and lowers latency.

3.7.5 PHY LAYER

This block has the responsibility of transmitting and reception of datagrams over a physical medium, which facilitates initiating and defusing a radio transceiver, energy recognition, joining eminence signal, frequency range, and the evaluation of the channel to the upper block. This block functions on a dual-band, which ranges from 2.4 gigahertz to 868–916 megahertz based on Direct Sequence Spread Spectrum [19]. Lesser propagation loss will occur when low transfer rate occurs at low band with larger range of functions. Apart from that a better frequency outcome in quicker amount of data, delay in time, and reduces burden [15].

3.8 IOT HARDWARE AND SENSORS

3.8.1 IoT HARDWARE

In the IoT, a hardware device is a physical device and mechanism designed to collect information, connect to devices, and communicate with devices connected via the Internet [20]. IoT hardware comes in various forms, sizes, and complexity levels, depending on the use case and application domain. As a result of IoT technologies, a growing number of sophisticated and diverse IoT hardware solutions have been developed [21].

The commonly used IoT hardware interfaces:

1. **Camera and Imaging Modules:** IoT devices are unified with cameras for object detection surveillance [22].
2. **Global Positioning System (GPS):** IoT devices may include different positioning modules, which provide information on locating objects over the devices through the Internet [23].
3. **Edge Computing Devices:** IoT applications are commonly used in edge computing devices to process the data locally instead of transferring all data to the cloud, which reduces latency and saves bandwidth [24].
4. **Display Interfaces:** Most IoT devices use display interfaces to communicate with the outsiders, such as Liquid Crystal Display (LCD), Light Emitting Diode (LED), and touchscreen displays to interact with the users [20].
5. **Gateways:** To maintain communication between IoT devices and other existing networks a gateway is a common technique used to collect information from multiple devices, develop the information and then send the information to other networks for further usage and storage.

6. **Connectivity Modules:** To exchange the processed information, IoT devices need to get connected to the Internet. Such communications can be established through cellular networks, Wi-Fi medium, Bluetooth, etc. [21].

7. **Actuators:** Based on the instructions received from the microcontrollers, allow the IoT devices to interact with the physical world.

8. **Sensors:** Sensors are the components that collect real-world information from the physical environment and detect the physical parameters such as temperature, humidity, light, motion, proximity, pressure, gas levels, etc., for each type of parameters different types of sensors are used based on the specific application [20].

9. **Microcontrollers and Microprocessors:** Microcontrollers and microprocessors are the Central Processing Unit (CPU) of IoT devices used to handle data processing, decision-making, and control of connected sensors and actuators.

Examples include Arduino, Raspberry Pi, etc. [24].

3.8.2   IoT Sensors

IoT sensors are a kind of hardware device which is used to collect information from outside the physical world or devices and translate that information into digital information which can be processed and communicated over the Internet [20].

These sensors are fundamental functioning unit of IoT systems helps in enabling various applications across different domains.

3.8.3   Common Types of IoT Sensors

Temperature sensors: which allows the users to scale the temperature, which can be used in applications such as climate control, weather monitoring, and industrial processes.

Humidity sensors: which helps in scaling the level of moisture in the air and are used in applications like Heating, Ventilation, and Air Conditioning (HVAC) systems, agriculture, and weather stations [21].

Light sensors (photodetectors): Detect the light levels in the physical world and are utilized by applications like automatic lighting control, display brightness adjustment, and outdoor lighting.

Motion sensors: Detect the movement or change in position and are used in security systems, smart lighting, and gaming applications [23].

Accelerometers: Scale the acceleration forces and are commonly used in fitness trackers, smartwatches, and motion-based applications.

Pressure sensors: Scale the atmospheric or fluid pressure and find applications in weather forecasting, altitude measurement, and industrial monitoring [24].

Gas sensors: Detect the existence of several gases in the atmosphere and are useful in quality monitoring and safety in industrial settings [23].

Sound sensors (microphones): Captures audio signals and is used in voice recognition noise monitoring.

Ultrasonic sensors: Use sound waves to detect the remoteness of an entity and are widely used in robotics, parking assistance systems, and industrial automation [21].

Image sensors (cameras): Capture the visual data and utilize that information in applications like surveillance systems, facial recognition, and object detection.

Stress sensors: Measure the stress and are used in health monitoring and industrial applications.
These are a few examples of the wide variety of IoT sensors available. Each type of sensor serves specific purposes in different IoT applications, making them valuable components in building smart and interconnected systems [25].

3.9 IOT WITH CLOUD COMPUTING (CC)

3.9.1 INTRODUCTION

IoT and CC are the two best examples that have gained attractions from the academics. Cloud computing provides services to users as an on-demand service that helps them avail the services from anywhere [26]. On the other hand, IoT uses hardware and sensors to provide service to the users as they smell changes in the physical world, and transmit information to the users who access it. We will discuss how both technologies work together to improve the betterment of the existing work [27].

3.9.2 FUSION OF IoT AND CC

IoT and CC have grown independently and created most of the impact in the technical field, but they are still reaching different levels. If the two technologies are combined, we can find huge benefits in the technical field.

Cloud computing is creating almost infinite possibilities in generating the huge amount of hardware and software like infrastructure, storage, and applications as a virtualized environment, which creates an advantage to IoT. When these cloud services and IoT devices come together, it creates an impact which can be considered as an evaluation [26].

As cloud computing is getting benefitted from IoT, IoT is also getting benefitted, such as endless volume and properties can be provided for the information which are collected by the IoT devices [28]. Cloud computing will significantly benefit from the widespread and advanced implementation of IoT, as it facilitates seamless communication between various devices in the physical environment [26]. By enabling this connectivity, cloud computing can provide innovative solutions and services tailored to a wide array of real-life scenarios, acting as a crucial intermediate block among things (devices) and requests.

Additionally, the connection between CC and the IoT creates matchless opportunities to collect, integrate, sharing data with others [29]. To ensure continuous availability and security, information from the cloud should be well-matched with a typical API, which permits it to benefit from healthy protective measures that are accessible from any location. Additionally, the dealing of IoT nodes doesn't necessarily need to occur within the devices themselves, as the data can be efficiently integrated and transferred to more capable storage nodes [30].

3.9.3 Services Provided

Merging of IoT and cloud services helps in developing functional services like:

- Sensing as a Service (SaaS)
- Database as a Service (DBaaS)
- Sensing and Actuation as a Service (SAaaS)
- A sensor as a Service (SenaaS)

SaaS: It is anticipated to build on top of the foundation of IoT structure and services to help in reaching the highest level of development and implementation. Using the SaaS cloud model, sensing servers can effectively manage and process sensing requests originating from various locations.

SAaaS: Cloud-based deployment of automated control logic is possible. For the deployment of SAaaS, sensors and actuators must be provided as reliable and well-defined services, bridging the gap regarding sensor networks and devices or compatible principles.

SenaaS: facilitates the all-encompassing management of distant sensors, allowing for their seamless and widespread control and monitoring.

DBaaS: It is a cloud-based, self-service method that enables users to deploy the collection of information and their applications directly to the intended recipients without requiring assistance from the processing department [26].

3.9.4 Benefits

Prior studies have demonstrated the broader benefits of combining IoT with cloud computing, encompassing advantages such as enhanced storage capacity, increased computational resources, and access to novel capabilities like advanced computing technologies and protocols. Additionally, it offers improved scalability, accuracy, and accessibility, which are typically challenging to obtain otherwise [31]. These benefits will prompt companies and individuals to contemplate the ways they can leverage the enabled technologies. This could aid organizations in addressing challenges related to cloud computing, such as limited financial resources and infrastructure, while also assisting them in making sense of their IoT information.

3.9.5 DEMERITS

The various challenges and complexities are.

   i. Software systems in smart gadgets.
   ii. Stabilization.
   iii. Establishing interoperability.

Concerns and challenges related to IoT cloud often led organizations to hesitate or be reluctant to adopt cloud technologies. Nevertheless, despite potential challenges and issues, it's essential to recognize that cloud computing also brings additional benefits [26].

3.10 EDGE COMPUTING AND IOT

3.10.1 INTRODUCTION

Using edge computing, advanced computing capabilities are brought closer to IoT nodes, facilitating the support of complex applications in the future IoT. By deploying powerful computing resources at the edge that are closer to users, edge computing offers real-time computing services [32]. Edge computing is used to decentralize computing power, moving it away from centralized cloud servers, and instead, placing it in edge nodes located near the end-users [33]. This approach yields two significant enhancements over traditional cloud computing. First, edge nodes can perform data preprocessing on a large scale before transmitting the data to central cloud servers.

The result is a reduction in the amount of data that needs to be transferred, resulting in an improved data handling process. Second, edge computing optimizes cloud resources by endowing edge nodes with their own computing capabilities [34].

3.10.2 EDGE COMPUTING AND IoT

An edge server is positioned at the network edge and connected directly to the access point (AP) of an Edge IoT network. There are numerous IoT nodes scattered throughout the area surrounding the access point. The edge server serves as a local computing hub, enabling real-time processing and analysis for the IoT devices and applications within its proximity [35].

These IoT nodes encompass various smart devices, sensors, actuators, or other objects equipped with embedded computing capabilities, forming an integral part of the IoT ecosystem. As these IoT nodes operate and generate data, the edge server plays a vital role in handling, aggregating, and responding to the information they produce [32].

3.10.3 EDGE COMPUTING IMPLEMENTATION

During the implementation of edge computing architecture, numerous research endeavors have been undertaken, resulting in several dominant models. These prevailing models have emerged as the most influential and widely adopted approaches in the field of edge computing [33].

1. Hierarchical model
2. Software-defined model

3.10.4 INTEGRATION OF IoT AND EDGE COMPUTING

Even though they operate independently, the integration of an edge computing platform can significantly assist IoT in resolving critical challenges and enhancing performance. By combining edge computing capabilities with IoT systems, various issues related to latency, bandwidth efficiency, data processing, and response times can be effectively addressed. Figure 3.2 shows the Layer Architecture of IoT-based Edge-Computing.

The likelihood of combining IoT with edge computing is with a particular emphasis on transmission, storage, and computation aspects. By narrowing our focus on these characteristics, we aim to demonstrate how edge computing can enhance the overall performance of IoT systems. Through this integration, IoT data transmission becomes more efficient, data storage and processing become faster and more responsive, leading to improved IoT performance and user experience.

Universally, IoT derives significant benefits from Edge and Cloud computing due to distinctive features offered by these dual constructions. Edge computing provides the advantage of high computational capacity and low-latency processing, thanks to its deployment closer to the IoT devices and sensors [36].

On the other hand, cloud computing offers extensive storage capabilities, allowing IoT data to be securely stored and easily accessible from anywhere. By harnessing the strengths of both edge and cloud computing, IoT systems can achieve a more robust and efficient ecosystem that optimizes both computational power and data storage.

Despite having inadequate computational dimensions and storing capacity, when compared to cloud computing, edge computing still offers distinct benefits for IoT. Additionally, edge computing excels in delivering fast response times, enabling real time meeting out and study of data directly by the side of the network's edge. This capability ensures that IoT applications can promptly respond to events and deliver timely actions [33].

3.10.5 ADVANTAGES

The integration of IoT with edge computing offers several advantages. By combining these two technologies, IoT systems can benefit from the reduction of latency, improved actual processing, improved information confidentiality and safety, optimized packet utilization, increased scalability, and more efficient use of computational resources [35]. As a result of edge computing, data can be processed and

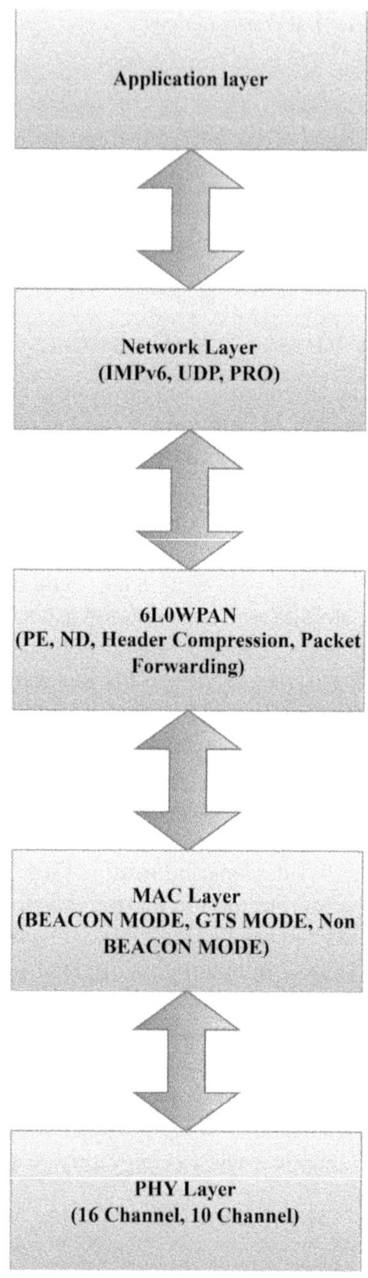

FIGURE 3.2 Layer architecture of IoT based edge-computing.

analyzed in a localized manner, allowing for faster response times and better overall performance [34].

1. Transmission
   a. Latency/delay, bandwidth, energy, overhead
2. Storage
   a. Storage balancing
   b. Recovery policy
      i. Availability
      ii. Data replication
3. Computation
   a. Computation offloading
      i. Local
      ii. Edge/cloudlet
      iii. Cloud
   b. Pricing policy
      i. Single service provider
      ii. Multiple service providers
   c. Priority

3.10.6 DISADVANTAGES

The integration of edge computing to support IoT brings forth numerous advantages. However, alongside these benefits, there are also challenges specific to edge computing-based IoT that need to be addressed [36]. These challenges may arise due to factors such as network connectivity issues, security concerns, scalability limitations, and the complexity of managing distributed computing resources at the edge. Edge Computing can significantly enhance IoT applications and services by identifying and overcoming these challenges [33].

1. System integration
2. Resource management
3. Security and privacy
4. Advanced communication
5. Smart system support

3.11 IOT DATA MANAGEMENT

3.11.1 INTRODUCTION

As a paradigm, the IoT has emerged in recent years, enabling billions of devices to communicate and generate vast quantities of data. Data management and analysis are essential for facilitating decision-making processes and gaining meaningful insights from them. As IoT-generated data is large, fast, varied, and verifiable, the study explores data management strategies tailored specifically for IoT environments. Sensors, actuators, and connected devices are all part of IoT data management, which involves gathering and integrating heterogeneous data from a variety of sources [37]. A network of interconnected devices, including phones, tablets, and wearables, the IoT communicates and exchanges data through the Internet. IoT has the potential to transform a variety of industries and aspects of our daily lives, in addition to enabling automation, improving efficiency, and facilitating intelligent decision-making. Data from sensors is used by edge gateway devices to make decisions through data analysis. Data must be collected in real time, organized, and communicated between these high-performance systems in real time. Data flow can be monitored by end users using web-based dashboards running on edge devices. When searching a large database containing a lot of data, graphs over arbitrary time periods can be used to locate records. It involves a variety of technologies and techniques, such as data warehouses, data lakes, and machine learning. IoT data management can be used for a variety of purposes, such as improving operational efficiency. IoT data can be used to identify inefficiencies in operations and make changes to improve efficiency. For example, IoT data can be used to track the performance of machines in a factory and identify machines that are not operating at peak efficiency. This information can be used to make changes to the production process, which can lead to increased productivity and decreased costs. Improving customer service: Customers' behavior and preferences can be tracked using IoT data. For example, IoT data can be used to track the browsing history of website visitors and recommend products that they are likely to find interesting. The result can be an increase in sales and a greater level of customer satisfaction for businesses. IoT data can be used to identify new market opportunities and to develop new products and services. For example, a supply chain can be tracked using data collected from IoT devices. In order to improve efficiency and lower costs, new logistics solutions can be developed based on this information. Automating tasks and improving efficiency can be achieved. An example of the use of IoT data is the tracking of inventory levels in a warehouse and the automatic ordering of additional inventory when levels decline [38].

3.11.2 Data Collection and Acquisition

Collecting and acquiring IoT data involves grouping data from IoT devices and making it available for analysis. Any IoT research must begin with this step, as it enables businesses to collect the data they need to make informed decisions and improve their operations. Decide what data will be collected? What data do you need to collect? This will depend on the specific application of your IoT system. Using IoT sensors, for example, to monitor the temperature of a manufacturing plant, you will need to collect temperature data at various points throughout the facility. Choose the right sensors. A wide variety of sensors are available, each with its strengths and weaknesses. You need to choose sensors that are accurate, reliable, and that can collect the data you need. Connect the sensors to the Internet. Once you've chosen the right sensors, you need to connect them to the Internet. In order to accomplish this, several technologies can be used, including Wi-Fi, Bluetooth, and cellular networks. A secure location must be selected to store the collected data once it has been collected. All can store the data on a local server, in the cloud, or in a combination of both. Many sensors are used to collect data from different environments. This will help us to collect data on a variety of factors, such as temperature, humidity, vibration, and motion. Store the data securely such as the data we collect should be stored securely in a location that is accessible to authorized users only. Make the most of your data by using data analytics tools. We should use data analytics tools to get the most out of our data. With the help of these tools, you are able to identify patterns and trends in your data, and then make informed decisions based on this information.

IoT data collection and acquisition: Increasingly, IoT devices are collecting large volumes of data. According to estimates, there will be 44.2 zettabytes of IoT data globally by 2020. By 2025, this number is expected to reach 181 zettabytes. The variety of data collected by IoT devices is also increasing. In addition to traditional data types such as temperature, humidity, and vibration, IoT devices are now collecting data on many factors, namely location, energy consumption, and customer behavior. The speed at which data is collected by IoT devices is also increasing. In the past, IoT devices would collect data at intervals of minutes, hours, or even days. Today, many IoT devices are collecting data in real-time. There are some emerging trends in IoT data collection and acquisition such as Edge Computing (EC), Cloud Computing (CC), Artificial Intelligence (AI), and Data Analytics (DA). There are various challenges faced during the data collection in IoT, such as compatibility with existing systems, huge amounts of data need to sort, real-time data need to stream. The recent research aims to bring a novel framework for IoT devices and a lightweight design to interact with various sensors in real-time data collection. For data transmission over the network a common protocol has been used such as HTTP, CoAP, and MQTT in the cloud storage [39].

3.11.3 Data Storage and Processing in IoT

The different use cases need to store the data in different formats; however, each device has the capacity to store data [39]. The worldwide is producing terabytes of information with various sources produced using IoT. Every generated data is essential for extracting useful information [40].

3.11.3.1 Cloud Edge Collaborative Storage (CECS)

The cloud server and IoT device are interconnected using CECS. CECS is clearly founded on the fact that cloud servers are wealthy in terms of capacity and computing assets, edge servers are situated near IoT devices, and IoT devices usually have limited assets [41]. Data will first be collected from IoT devices and sent to nearby edges, where they will be processed in real time, returned results, and stored on cloud servers. The user can ultimately share the IoT data expected on the cloud server by creating relevant search queries.

3.11.3.2 P2P File System Based on Blockchain Technology

The framework for verifying the integrity of large-scale IoT data using blockchain technology is shown in Figure 3.3. It is recommended that IoT devices sign blockchain transactions or that other nodes verify the device's signature. IoT devices' flaws can be fixed by using edge servers. Several IoT devices near the edge server constantly

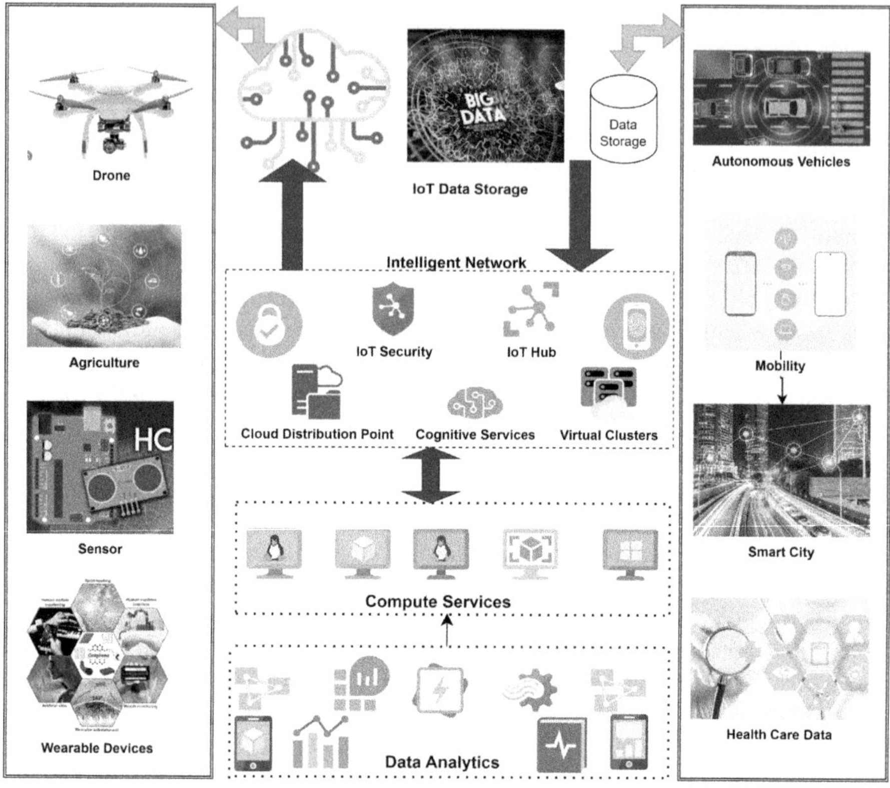

FIGURE 3.3 Internet of things (IoT) general solution architecture and storage.

collect data [42]. In this case, it determines which addresses account for cloud servers that will be used for storing data and sends blocks of data to those hosts. An edge server maintains a copy of each device's identity and helps each device generate homomorphic verifiable tags and data shards. In the prototype system, edge computing processors are located close to the IoT devices in order to preprocess large amounts of IoT data. This has the potential to significantly reduce the communication costs and computational load associated with the system. Traditionally, P2P solutions were based on client-server architecture where the server had powerful processing abilities.

3.11.3.3 Big Data Storage

IoT devices can generate massive amounts of data, and Big Data technologies are well-suited for handling and processing such large-scale data. A distributed and scalable data storage system is typically used to store IoT data in Big Data environments. Distributed file systems like Hadoop Distributed File System (HDFS) are designed to save huge amounts of data on a number of servers. HDFS provides fault tolerance, data replication, and high throughput, making it suitable for IoT data storage. NoSQL Databases: Various NoSQL databases like Apache Cassandra, MongoDB, or Apache

HBase are commonly used for IoT data storage. In addition to handling unstructured and semi-structured data, NoSQL databases also handle IoT data [43].

3.11.4   DATA ANALYTICS AND INSIGHTS IN IoT

A data analytics approach involves analyzing both quantitative and qualitative data about a digital property, in order to draw meaningful conclusions from them. Sensor/ Textual data produce a huge number of real-time datasets. As a result of the lack of decentralized control, the management of IoT-based sensors is a complex process, and aggregating data can be difficult. Due to their dynamic properties and limited computational power, IoT sensor systems can benefit from using swarm intelligence to solve complex problems. These datasets are preprocessed to utilize deep learning architecture and swarm intelligence for analytics. In addition to providing a complex problem solution using GPU-based memory, Tata Consultancy Service (TCS) performs parallel analyzes of streaming data from multiple sensors 200 times faster than traditional methods [44]. It processes huge amounts of sensor data and generates useful insights. Social media data analytics utilize natural language processing and text mining in order to extract useful knowledge. Convolutional neural networks and video content analysis were used to represent useful data from image/video data. Through spatial, temporal visualization and location intelligence, geospatial data analytics can help us explore high-resolution images and data. A geospatial intelligence application can provide unique insights, reveal hidden patterns, and deliver information for better decision-making based on geospatial data [45]. Analyzing all of these data sets in an automated manner will be possible with data analytics.

3.11.5   SECURITY AND PRIVACY IN IoT

Digital technologies, namely artificial intelligence, blockchain, and machine learning have been used to improve the security of data collected at the device level. IoT device data currently is not protected by current standards, resulting in individual confidentiality issues [46]. There are many inconsistencies and gaps in the current standards for securing IoT data, and they contribute to inconsistencies and gaps in future standards. Examples of typical dataflow management applications in smart manufacturing, smart transportation, and smart cities are presented to illustrate the practical benefits of efficient IoT dataflow management [47]. Transporting encrypted data at rest or in motion using standards such as Secure Socket Layer (SSL), Federal Information Protection Standard (FIPS), Secure Socket Shell (SSH), and ISO 27001 is used in communication middleware. IoT devices have a few basic level security measures based on role-based access control, encryption with regular updates and patches, and granular permissions. The development of smart and intelligent sensors and actuators, as well as RFID tags, has forward to the creation of a huge number of wireless networks with smart and intelligent equipment (things) connected to the Internet constantly transmitting data. The IoT makes it extremely challenging to maintain privacy and security for all this data. Therefore, numerous present and future applications must prioritize safety and security.

3.11.5.1 Introduction

Networking and connectivity are fundamental pillars of the IoT ecosystem. The IoT devices are constructed on a system with interconnected gadgets that can network and share data instantly. IoT networking and connectivity encompass a wide range of technologies, protocols, and architectures that enable efficient and secure communication among IoT devices and cloud-based services. Networking refers to the integration of computing devices to share resources, data, and information. It allows computers, servers, printers, and other devices to communicate locally and internationally (through the Internet). Networking is an essential component of modern technology, facilitating communication, and data transfer between devices.

Connectivity in IoT is a vast field with various subdisciplines, including network administration, network security, network design, and network engineering. For various types of networking, connectivity is also used in the IOT application. The IoT relies heavily on networking to enable communication and data exchange between a vast number of connected devices. IoT refers to a concept in which everyday objects or devices are arranged through sensors, actuators, and communications capabilities, allowing them to collect and exchange data via the Internet or other networks. From smart home appliances to wearable technology to industrial machinery and smart city infrastructure, these interconnected devices can serve a variety of purposes.

In today's interconnected world, it is critical for enterprises, organizations, and individuals to understand networking fundamentals in order to facilitate efficient and secure communication. Typically, network protocols describe how data is transmitted over a computer network according to a set of rules and conventions. Data packet protocols define not only the format and structure of data packets, but also the procedures involved in creating, maintaining, and terminating communication sessions between devices connected to a network, as well as the structure of the data packets themselves. These are the physical devices used in a network, such as computers, routers, switches, access points, and servers.

There are several key components and concepts in networking devices:

- **Protocols**: Transmitting and receiving data is controlled by protocols, which are rules that govern the transmission and reception of data. The purpose of a network protocol is to facilitate communication between computers that are connected to a local area network (LAN), a wide area network (WAN), or the Internet. They ensure that devices from different manufacturers and running different software can communicate effectively and reliably.
    a. **LAN**: In computer networking, a LAN is a network of devices that are interconnected within a small area, usually within a building or a household. The Internet, printers, and files are commonly shared over LANs.
    b. **WAN**: The WAN connects LANs and other networks across long distances and geographically dispersed areas. WANs are most commonly associated with the Internet.

- **Router**: A router is an electronic device that connects several networks and sends data packets between them. To transport data from a source to a destination, it determines the best route.
- **Switch**: LAN switches are devices used to connect multiple devices. It uses MAC addresses to determine where to forward data within the local network.
- **Firewall**: According to predefined security rules, a firewall analyzes and prevents network traffic. As a result, unauthorized access and other threats are prevented from entering the network.
- **Wireless Networking**: Using radio waves, wireless networks allow devices to be connected without having to rely on physical cables to do so. Wi-Fi exemplifies wireless networking.
- **Internet Protocol (IP) Address**: Protocols such as this one are responsible for routing and addressing data packets in order for them to be properly delivered across different networks. An Internet Protocol address is a number designation that is unique to each device on a network. It enables devices to connect and interact with one another via an IP-based network.
- **Domain Name System (DNS)**: A domain name is converted into an IP address that computers can use to find each other on the Internet by converting human-readable characters into numbers.
- A DNS is a protocol for translating human domain names into IP addresses. Rather than entering a numerical IP address, users can access websites using easy-to-remember names.

Networking in IoT is a multidimensional challenge that requires considerations for scalability, security, energy efficiency, and interoperability to create robust and reliable ecosystems of connected devices.

3.11.6 WIRELESS TECHNOLOGIES FOR IoT

Wireless technologies enable seamless communication between devices without physical connections, which is essential for the widespread adoption of IoT applications. As technology continues to advance, IoT networking will continue to evolve to meet the demands of various industries and use cases. IoT devices utilize various communication protocols to establish connections and exchange data [48]. There are several wireless technologies specifically designed to meet the requirements of IoT devices, each with its strengths and use cases. Here are some of the prominent wireless technologies used in IoT.

Wireless Connectivity: Most IoT devices utilize wireless communication technologies due to their flexibility and ease of deployment. Some common wireless connectivity options in IoT include:

Wi-Fi: Wireless technology such as Wi-Fi allows fast data transfer between devices and the Internet, making it ideal for applications requiring fast communication. Data can be transferred over LANs at high speeds, making it suitable for home, office, and industrial settings where Wi-Fi access points are nearby.

Bluetooth: Designed for short-range communications between devices with low power consumption, Bluetooth is the ideal technology for this application. A wide variety of wearable devices, smart home gadgets, and health monitoring applications make use of this technology.

Zigbee: Wireless communication protocols such as Zigbee are designed for devices to communicate at low power, low rate, and short distances. This technology is primarily used to create wireless sensor and control networks, also called the "Internet of Things" (IoT). As part of IEEE 802.15.4, Zigbee provides a framework for creating reliable and scalable wireless networks. The system is generally used in smart lighting applications, industrial monitoring applications, and home automation applications.

Z-Wave: In addition, Z-Wave is another wireless protocol that is capable of supporting home automation and smart homes. Unlike other smart home networks, it operates in the sub-GHz frequency range, which provides good range and reliability. This type of network protocol is commonly used for smart home applications, particularly for home automation and security systems.

Cellular Connectivity: IoT devices can leverage existing cellular networks (2G, 3G, 4G, and now transitioning to 5G) to connect to the Internet over long distances. Cellular connectivity is particularly suitable for IoT deployments in remote areas or applications requiring wide-area coverage, such as asset tracking and smart city solutions.

Low-Power WAN (LPWAN): This technique caters specifically to IoT devices that require long-range connectivity while operating on low power. It operates on a licensed spectrum, offering better security and reliability compared to some unlicensed LPWAN technologies.
    Examples include:

- **LoRaWAN**: The technology is ideal for applications such as smart cities, agriculture, and environmental monitoring due to its long range and low power consumption.
- **NB-IoT (Narrowband IoT)**: IoT deployments requiring low-power and wide-area communication using cellular-based LPWANs.
- **Sigfox**: Sigfox is an LPWAN technology that operates in the unlicensed spectrum and is known for its simplicity and low-cost connectivity. It is suitable for applications such as asset tracking, environmental monitoring, and smart city management. A proprietary LPWAN technology known for its simplicity and low-energy requirements.

Wired Connectivity: Though less common in many consumer IoT applications, some devices may utilize wired connectivity, such as Ethernet, to connect to the Internet or local networks. This approach is more prevalent in IIoT settings.

Mesh Networking: Mesh networks are formed when IoT devices communicate with each other directly, rather than relying solely on a centralized hub or access point. Mesh networking is advantageous for its self-healing capabilities and extended coverage range. Zigbee and Thread are examples of protocols used to create mesh networks.

Edge Computing: IoT generates a large amount of data, which is difficult to transfer efficiently to centralized servers due to the large amount of data generated. In order to conserve network bandwidth and reduce latency, edge computing is used to process and analyze data at the edge (i.e., the IoT device) in order to process and analyze data.

Security: Security concerns arise from IoT connectivity. Implementing secure authentication, encryption, and data integrity measures on devices can prevent cyber threats and unauthorized access to sensitive data.

Interoperability: A cohesive IoT ecosystem requires IoT devices to be able to communicate despite differences in manufacturers, platforms, or protocols. Standardization efforts and common communication protocols contribute to achieving interoperability.

Communication range, power requirements, data rate, deployment environment, and cost are factors that must be considered when selecting a wireless technology for an IoT application. As IoT continues to evolve, new wireless technologies and improvements in existing ones are likely to emerge, further expanding the possibilities for IoT deployments.

3.11.7 Cellular Networks and IoT

IoT devices need cellular networks to communicate and connect across large geographical areas. Suitable for long-range communications and seamless connectivity, these devices deliver reliable, secure, and high-speed data transmission. It involves the use of the same technology that powers your smartphone to connect physical objects (such as sensors) to the Internet. IoT devices do not require a separate, private network, but can instead be connected to the same mobile network as smartphones. There is a possibility that cellular IoT is able to provide an alternative to non-cellular LPWAN, like LoRaWAN or Sigfox.

Cellular networks running on 4G provide a broadband connection for voice calls and video streaming applications in the consumer mobile markets. Yet, they are pretty costly. They demand a lot of electricity, and it can be difficult to provide partial inside coverage in industrial settings. They do, however, make a fantastic choice for LPWAN networks. Ultimately, IoT data must be sent through the Internet, with 4G being the best global choice for capacity, scalability, and compatibility [49]. Recent news coverage of 5G has been mixed, with positive and negative aspects. The technology, supports mobility at extreme speeds with extremely low latency. Self-driving cars and augmented reality are expected to use this technology in the future.

Wide Area Coverage is one of the primary advantages of using cellular networks for IoT is their extensive coverage. Cellular infrastructure is already well-established in many regions, providing connectivity even in remote areas. This wide area coverage allows IoT devices to operate in various locations, making them suitable for applications such as environmental monitoring, smart agriculture, and asset tracking.

- **High Data Transfer Rates**: In the last few years, cellular technologies have dramatically improved the transfer of data from 3G to 4G and now to 5G. This improvement in bandwidth enables real-time communication and high-throughput applications, making cellular networks suitable for video surveillance, industrial automation, and healthcare monitoring.
- **Reliability and QoS**: Cellular networks are designed to offer high reliability and consistent connectivity. Unlike smart grids, transportation systems, and emergency services, which are dependent on IoT, this is essential for mission-critical IoT applications, where downtime or communication interruptions can have severe consequences.
- **Security**: A cellular network protects the data that is transmitted between IoT devices and the cloud, including encryption and authentication, as well as other security components. This level of security is critical for safeguarding sensitive information and preventing unauthorized access to IoT systems.
- **Global Roaming**: For IoT applications with international deployments, cellular networks support global roaming capabilities. This allows devices to connect seamlessly to different networks in various countries, enabling international tracking, logistics, and asset management solutions.
- **Cost-Efficiency**: With the widespread adoption of cellular technologies, the cost of connectivity has become more affordable. Many cellular network providers offer cost-effective data plans and packages tailored for IoT deployments, making it more accessible for businesses and organizations to implement large-scale IoT solutions.
- **Mobility Support**: Cellular networks support mobility, allowing IoT devices like vehicles and wearable devices to maintain continuous connectivity while on the move. This capability is essential for applications like connected cars, fleet management, and healthcare wearables.
- **Future-Proofing with 5G**: The emergence of 5G technology brings significant enhancements to cellular networks, such as lower latency, higher data rates, increased device density, and support for massive machine-type communication (mMTC). 5G's attributes will unlock new possibilities for IoT applications, particularly in areas requiring real-time responsiveness, like augmented reality, industrial automation, and smart cities.

It was recommended for submission for consideration as an International Mobile Telecommunications (IMT)-2020 standard by the wireless industry standardization body 3rd Generation Partnership Project (3GPP), and it is capable of delivering significant increases in gigabit rates for both uploads and downloads. Despite the advantages, cellular networks do have some limitations for certain IoT use cases. The reliance on cellular infrastructure can result in higher power consumption for IoT devices, which may be a concern for battery-operated devices that require long-term deployments [50]. Also, LPWAN or satellite communications might be a better option in areas with low cellular coverage. Overall, cellular networks remain a significant enabler for IoT, offering reliable, scalable, and secure connectivity for a wide range of submissions across businesses. Mobile networks will continue to play an increasingly important role in supporting IoT ecosystems as technology evolves.

3.11.8 SHORT-RANGE COMMUNICATION IN IoT

IoT devices can exchange data over relatively close distances with each other and with other IoT devices using short-range communications. These technologies are particularly useful in scenarios where devices need to interact within a confined space or when they are in proximity to each other. Short-range communication technologies are especially useful in scenarios where long-range connectivity is unnecessary, impractical, or consumes too much power. These devices are primarily used in buildings, homes, factories, and other environments in which devices must interact closely with one another.

One of the significant advantages of short-range communication is its lower power consumption, which allows for prolonged battery life in IoT devices. Short-range technologies often have lower implementation costs, making them more accessible for various IoT deployments. In order to achieve the greatest level of IoT integration with other devices and platforms, it's important to select the right technology for short-range data transfer [51]. As the IoT ecosystem continues to evolve, short-range communication technologies will remain vital components, complementing long-range connectivity options and driving innovation across various industries.

When using RFID, which transmits small quantities of data, a reader and an RFID tag can be separated by just a few centimeters. Technology has been used in the retail and logistics sectors. A variety of items and machinery can be embedded with RFID tags to track inventory and assets in real-time. Improved supply chain management and better stock planning are possible as a result. Self-checkout, smart mirrors, and other IoT applications are made possible by RFID's continued entrenchment in the retail industry, which is growing IoT adoption.

3.11.9 IoT GATEWAYS AND EDGE DEVICES

As an intermediary between IoT devices deployed in the field and the cloud, an IoT gateway is responsible for bridging the gap between these two environments. It connects various sensors and devices to the cloud for processing and analysis after the data has been aggregated. An IoT architecture cannot be completed without gateways and edge devices. These platforms facilitate efficient communication, data processing, and connectivity between IoT devices and cloud-based platforms. In the context of the IoT, it is important to keep in mind that many devices use different communication protocols and technologies. As a result of translating and bridging these protocols, the gateway is able to ensure seamless communication and data exchange across heterogeneous devices and the cloud. Figure 3.4 mentions the different IoT gateways, which come in a variety of shapes, sizes, and types.

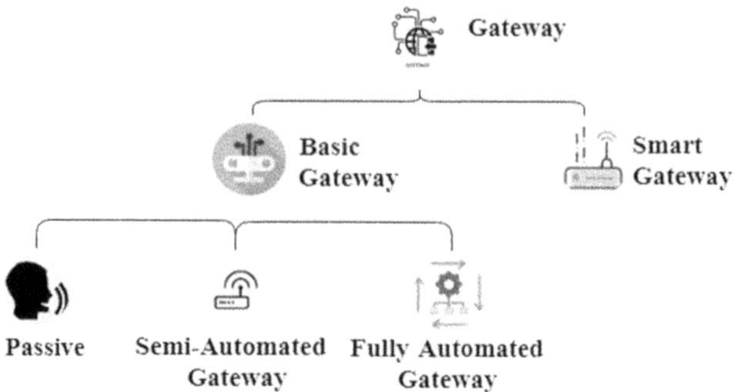

FIGURE 3.4 IoT gateways come in a variety of shapes, sizes, and types.

The success of IoT implementations depends on both IoT gateways and edge devices. The use of IoT gateways facilitates communication between devices and the cloud by facilitating data flow between them [52]. Edge devices, on the other hand, serve as the frontline data collectors and processors, bringing real-time capabilities and localized intelligence to the IoT ecosystem. Together, they form a robust and responsive IoT infrastructure that addresses the unique challenges and requirements of various IoT applications.

In addition to interconnecting sensing networks, gateways are used to connect cloud computing environments or data centers to the Internet. Its objective is to manage heterogeneous data collected by multiple devices in many forms and deliver this data to a developed phase [53]. Before transferring assembled data to data centers for optimal operation and management of IoT devices, it should be cleaned, pre-processed, and filtered. In terms of functionality, gateways can be divided into two types: basic gateways and smart gates. By forwarding data incoming from low-end IoT devices to data centers, the basic gateway serves as a gateway between them. In contrast, a smart gateway performs data processing efficiently by preprocessing, filtering, and analyzing the data prior to distributing it to the cloud database. The purpose of these intermediary gadgets is to function in harsh environments and improve from failures while restoring the displacement gap in the shortest amount of time possible.

As a gateway, software, and hardware equipment are used together. It consists of a small software package that manages, preprocesses, and stores nodes in a network. Additionally, a modest backup should be included to handle the connection defeat issue, which involves maintaining the current configuration, running it in nested mode for restoration, and performing data restoration from the same location.

3.11.10 IoT Network Protocols and Standards

Devices, gateways, and cloud platforms must work together in order to communicate efficiently. An IoT network protocol and standard facilitate this exchange of data and interoperability. These protocols and standards ensure that IoT systems can work seamlessly together, regardless of the manufacturers or technologies involved.

Some of the real IOT application information needs to be exchanged in the various protocols, making it popular for M2M communication. A lightweight MQTT, publish-subscribe messaging protocol caters to low-bandwidth, high-latency, and unreliable networks [54]. In order to operate on devices and networks with limited resources, CoAP is developed to be a lightweight application layer protocol. As part of the IoT, it is commonly used in applications where devices have limited processing power and memory, namely IoT smart homes and IoT industrial automation systems. Among IoT devices and cloud-based platforms, HTTP is widely used for communication. As IoT devices are able to communicate with RESTful APIs via HTTP, they can be easily added to existing web applications and web services. A Bluetooth Low Energy (BLE) protocol is a wireless communication protocol that is widely used in IoT, especially in wearable devices, healthcare applications, and smart home systems. IoT mesh networks can be created using ZigBee devices, which provide low-power, low-data-rate wireless communication, particularly for industrial and smart home applications. Device management protocols such as LwM2M are designed to help IoT devices and applications to be managed efficiently.

As new technologies and innovations emerge, the IoT network protocol and standard landscape continue to evolve. It is often difficult to determine which protocol is most appropriate depending on the specific IoT use case, device constraints, data transfer requirements, and the current ecosystem of platforms and devices [55]. To facilitate seamless communication between IoT devices and services, unified standards and protocols are being developed to facilitate interoperability between different protocols and standards.

REFERENCES

1. Jianxin Wang, Ming K. Lim, Chao Wang and Ming-Lang Tseng, "The Evolution of the Internet of Things (IoT) over the Past 20 Years," *Computers & Industrial Engineering*, vol. 155, p. 107174, 2021, ISSN 0360-8352, https://doi.org/10.1016/j.cie.2021.107174.

2. Anass Sedrati, Abdellatif Mezrioui and Aafaf Ouaddah, "IoT-Gov: A Structured Framework for Internet of Things Governance," *Computer Networks*, vol. 233, p. 109902, 2023, ISSN 1389-1286, https://doi.org/10.1016/j.comnet.2023.109902.

3. Čolaković A., Hadžialić, M., "Internet of things (IoT): A Review of Enabling Technologies, Challenges, and Open Research Issues," *Computer Networks*, vol. 144, pp. 17–39, Oct. 2018, https://doi.org/10.1016/j.comnet.2018.07.017.

4. Yosra Hajjaji, Wadii Boulila, Imed Riadh Farah, Imed Romdhani and Amir Hussain, "Big Data and IoT-based Applications in Smart Environments: A Systematic Review," *Computer Science Review*, vol. 39, p. 100318, 2021, ISSN 1574-0137, https://doi.org/10.1016/j.cosrev.2020.100318.

5. M. Z. Khan, O. H. Alhazmi, M. A. Javed, H. Ghandorh and K. S. Aloufi, "Reliable Internet of Things: Challenges and Future Trends," *Electronics*, vol. 10, no.19, pp. 23–77, Sep. 2021, https://doi.org/10.3390/electronics10192377.

6. Kehinde Lawal and Hamed Nabizadeh Rafsanjani, "Trends, Benefits, Risks, and Challenges of IoT Implementation in Residential and Commercial Buildings," *Energy and Built Environment*, vol. 3, no. 3, pp. 251–266, 2022, ISSN 2666-1233, https://doi.org/10.1016/j.enbenv.2021.01.009.

7. Ibrar Yaqoob, Khaled Salah, Raja Jayaraman and Mohammed Omar, "Metaverse Applications in Smart Cities: Enabling Technologies, Opportunities, Challenges, and

Future Directions," *Internet of Things*, vol. 11, p. 100884, 2023, ISSN 2542-6605, https://doi.org/10.1016/j.iot.2023.100884.

8. Mirani, G. Velasco-Hernandez, A. Awasthi and J. Walsh, "Key Challenges and Emerging Technologies in Industrial IoT Architectures: A Review," *Sensors*, vol. 22, no. 15, p. 5836, Aug. 2022, https://doi.org/10.3390/s22155836.

9. Nishant Chaurasia and Prashant Kumar, "A Comprehensive Study on Issues and Challenges Related to Privacy and Security in IoT," *e-Prime – Advances in Electrical Engineering, Electronics and Energy*, vol. 4, p. 100158, 2023, ISSN 2772-6711, https://doi.org/10.1016/j.prime.2023.100158.

10. *Evolution of Internet of Things.* www.techaheadcorp.com/knowledge-center/evolution-of-iot/, accessed 03 August 2023.

11. Security, privacy and trust of different layers in Internet-of-Things (IoTs) framework by A. Tewari and B.B. Gupta.

12. M. M. Islam, S. Nooruddin, and F. Karray, "Internet of Things: Device Capabilities, Architectures, Protocols, and Smart Applications in Healthcare Domain," *IEEE Internet of Things Journal*, vol. 7, no. 10, pp. 10093–10108, Oct. 2020.

13. H. A. Khattak, M. A. Shah, S. Khan, I. Ali, and M. Imran, "Perception Layer Security in Internet of Things," *Future Generation Computer Systems*, vol. 100, pp. 144–164, Nov. 2019.

14. M. Lombardi, F. Pascale, and D. Santaniello, "Internet of Things: A General Overview between Architectures, Protocols and Applications," *Information Systems Frontiers*, vol. 23, pp. 1261–1281, Mar. 2021.

15. B. Padma and S. B. Erukala, "Peer-to-Peer Communication Protocol in IoT-Enabled ZigBee Network: Investigation and Performance Analysis," *Journal of Ambient Intelligence and Humanized Computing*, vol. 12, pp. 7497–7505, 2021.

16. C. K. Rath, A. K. Mandal, and A. Sarkar, "Microservice Based Scalable IoT Architecture for Device Interoperability," in *Proceedings of the 2020 IEEE International Conference on Advanced Networks and Telecommunications Systems (ANTS)*, 2020, pp. 1–6.

17. A. Gerodimos, L. Maglaras, M. A. Ferrag, N. Ayres, and I. Kantzavelou, "IoT: Communication Protocols and Security Threats," in *Proceedings of the 2019 IEEE Global Communications Conference (GLOBECOM)*, 2019, pp. 1–6.

18. R. Attarian and S. Hashem, "An Anonymity Communication Protocol for Security and Privacy of Clients in IoT-Based Mobile Health Transactions," *IEEE Transactions on Industrial Informatics*, vol. 16, no. 4, pp. 2702–2711, Apr. 2020.

19. C. Bayılmış, M. A. Ebleme, Ü. Çavuşoğlu, K. Küçük, and A. Sevina, "A Survey on Communication Protocols and Performance Evaluations for Internet of Things," *Internet of Things*, vol. 9, p. 100149, Sep. 2020.

20. T. Domínguez-Bolaño, O. Campos, V. Barral, C. J. Escudero, and J. A. García-Naya, "An Overview of IoT Architectures, Technologies, and Existing Open-Source Projects," *Electronics*, vol. 9, no. 2, p. 275, Feb. 2020.

21. J. Trevathan, W. Read, and A. Sattar, "Implementation and Calibration of an IoT Light Attenuation Turbidity Sensor," *Sensors*, vol. 19, no. 9, p. 2130, May 2019.

22. J. D. A. Correa, A. S. R. Pinto, and C. Montez, "Lossy Data Compression for IoT Sensors: A Review," *IEEE Access*, vol. 8, pp. 68506–68527, 2020.

23. M. Kumar, "A Survey on Event Detection Approaches for Sensor Based IoT," *Journal of King Saud University-Computer and Information Sciences*, vol. 33, no. 5, pp. 489–500, Jun. 2021.

24. F. Famá, J. N. Faria, and D. Portugal, "An IoT-Based Interoperable Architecture for Wireless Biomonitoring of Patients with Sensor Patches," *Sensors*, vol. 20, no. 18, p. 5266, Sep. 2020.

25. I. Nassra and J. V. Capella, "Data Compression Techniques in IoT-Enabled Wireless Body Sensor Networks: A Systematic Literature Review and Research Trends for QoS Improvement," *IEEE Access*, vol. 8, pp. 61980–62000, 2020.

26. W. N. Hussein, "A Proposed Framework for Healthcare Based on Cloud Computing and IoT Applications," *Journal of King Saud University-Computer and Information Sciences*, vol. 33, no. 7, pp. 918–924, Aug. 2021.

27. A. Rahman, M. J. Islam, S. S. G. Muhammad, K. Hasan, and P. Tiwari, "Towards a Blockchain-SDN-Based Secure Architecture for Cloud Computing in Smart Industrial IoT," *IEEE Internet of Things Journal*, vol. 8, no. 4, pp. 2336–2344, Feb. 2021.

28. C. Yang and H. Ming, "Detection of Sports Energy Consumption Based on IoTs and Cloud Computing," *Computers, Materials & Continua*, vol. 68, no. 3, pp. 3177–3190, 2021.

29. R. Bose, "Design of Smart Inventory Management System for Construction Sector Based on IoT and Cloud Computing," *Wireless Personal Communications*, vol. 118, no. 1, pp. 541–557, May 2021.

30. G. Peng, "Constrained Multi Objective Optimization for IoT-Enabled Computation Offloading in Collaborative Edge and Cloud Computing," *Computers & Electrical Engineering*, vol. 93, p. 107260, Oct. 2021.

31. M. Wang and Q. Zhang, "Optimized Data Storage Algorithm of IoT Based on Cloud Computing in Distributed System," *IEEE Access*, vol. 8, pp. 196695–196705, 2020.

32. S. Shen, "Optimal Privacy Preservation Strategies with Signalling Q-learning for Edge-Computing-Based IoT Resource Grant Systems," *Computers & Electrical Engineering*, vol. 92, p. 107168, Sep. 2021.

33. J. Li, "Service Home Identification of Multiple-Source IoT Applications in Edge Computing," *Sensors*, vol. 21, no. 12, p. 4125, Jun. 2021.

34. M. Goudarzi, "An Application Placement Technique for Concurrent IoT Applications in Edge and Fog Computing Environments," *Journal of Network and Computer Applications*, vol. 182, p. 103016, Dec. 2021.

35. Y. Chen, "Channel-Reserved Medium Access Control for Edge Computing Based IoT," *IEEE Access*, vol. 8, pp. 219297–219305, 2020.

36. W. Yu, "A Survey on the Edge Computing for the Internet of Thing," *IEEE Access*, vol. 8, pp. 102161–102174, 2020.

37. S. Chen, X. Zhu, H. Zhang, C. Zhao, G. Yang and K. Wang, "Efficient Privacy Preserving Data Collection and Computation Offloading for Fog-Assisted IoT," *IEEE Transactions on Sustainable Computing*, vol. 5, no. 4, pp. 526–540, 1 Oct.–Dec. 2020, https://doi.org/10.1109/TSUSC.2020.2968589.

38. P. Ta-Shma, A. Akbar, G. Gerson-Golan, G. Hadash, F. Carrez and K. Moessner, "An Ingestion and Analytics Architecture for IoT Applied to Smart City Use Cases," *IEEE Internet of Things Journal*, vol. 5, no. 2, pp. 765–774, Apr. 2018, https://doi.org/10.1109/JIOT.2017.2722378.

39. J.-S. Fu, Y. Liu, H.-C. Chao, B. K. Bhargava and Z.-J. Zhang, "Secure Data Storage and Searching for Industrial IoT by Integrating Fog Computing and Cloud Computing," *IEEE Transactions on Industrial Informatics*, vol. 14, no. 10, pp. 4519–4528, Oct. 2018, https://doi.org/10.1109/TII.2018.2793350.

40. Q.-T. Doan, A. S. M. Kayes, W. Rahayu and K. Nguyen, "Integration of IoT Streaming Data With Efficient Indexing and Storage Optimization," *IEEE Access*, vol. 8, pp. 47456–47467, 2020, https://doi.org/10.1109/ACCESS.2020.2980006.

41. Y. Tao, P. Xu and H. Jin, "Secure Data Sharing and Search for Cloud-Edge-Collaborative Storage," *IEEE Access*, vol. 8, pp. 15963–15972, 2020, https://doi.org/10.1109/ACCESS.2019.2962600.

42. H. Wang and J. Zhang, "Blockchain Based Data Integrity Verification for Large-Scale IoT Data," *IEEE Access*, vol. 7, pp. 164996–165006, 2019, https://doi.org/10.1109/ACCESS.2019.2952635.

43. M. Babar, M. A. Jan, X. He, M. U. Tariq, S. Mastorakis and R. Alturki, "An Optimized IoT-Enabled Big Data Analytics Architecture for Edge–Cloud Computing," *IEEE Internet of Things Journal*, vol. 10, no. 5, pp. 3995–4005, 1 Mar. 2023, https://doi.org/10.1109/JIOT.2022.3157552.

44. A. R. Dargazany, P. Stegagno, and K. Mankodiya, "WearableDL: Wearable Internet-of-Things and Deep Learning for Big Data Analytics—Concept, Literature, and Future," *Mobile Information Systems*, no. 1, 8125126.

45. S. A. Shah, D. Z. Seker, S. Hameed and D. Draheim, "The Rising Role of Big Data Analytics and IoT in Disaster Management: Recent Advances, Taxonomy and Prospects," *IEEE Access*, vol. 7, pp. 54595–54614, 2019, https://doi.org/10.1109/ACCESS.2019.2913340.

46. D. Wei et al., "Dataflow Management in the Internet of Things: Sensing, Control, and Security," *Tsinghua Science and Technology*, vol. 26, no. 6, pp. 918–930, Dec. 2021, https://doi.org/10.26599/TST.2021.9010029.

47. J. Cook, S. U. Rehman and M. A. Khan, "Security and Privacy for Low Power IoT Devices on 5G and Beyond Networks: Challenges and Future Directions," *IEEE Access*, vol. 11, pp. 39295–39317, 2023, https://doi.org/10.1109/ACCESS.2023.3268064.

48. Gunjan Beniwal and Anita Singhrova, "A Systematic Literature Review on IoT Gateways", *Journal of King Saud University – Computer and Information Sciences*, vol. 34, no. 10, Part B, 2022, pp. 9541–9563, ISSN 1319–1578, https://doi.org/10.1016/j.jksuci.2021.11.007.

49. Vaigandla, Karthik, Radha, Krishna and Allanki, Sanyasi Rao, A Study on IoT Technologies, Standards and Protocols. 2021. https://doi.org/10.17697/ibmrd/2021/v10i2/166798.

50. Pons, Mario, Estuardo Valenzuela, Brandon Rodríguez, Juan Arturo Nolazco-Flores and Carolina Del-Valle-Soto, "Utilization of 5G Technologies in IoT Applications: Current Limitations by Interference and Network Optimization Difficulties—A Review", *Sensors*, vol. 23, no. 8, p. 3876, 2023, https://doi.org/10.3390/s23083876.

51. Tariq, Usman, Irfan Ahmed, Ali Kashif Bashir, and Kamran Shaukat, "A Critical Cybersecurity Analysis and Future Research Directions for the Internet of Things: A Comprehensive Review", *Sensors*, vol. 23, no. 8, p. 4117, 2023, https://doi.org/10.3390/s23084117.

52. Yaser Mohammed Al-Worafi, *"Chapter 6 Internet Technologies for Drug Safety"*, Springer Science and Business Media LLC, 2023. https://doi.org/10.1007/978-3-031-34268-4_6.

53. Isa Avci and Cevat Özarpa. *"Chapter 10 Machine Learning Applications and Security Analysis in Smart Cities"*, Springer Science and Business Media LLC, 2022. https://doi.org/10.1007/978-3-030-97516-6_10.

54. Laguidi, Ahmed, Tamtam, Samiya and Mejdoub, Youssef. "A Technique to Improve IoT Connectivity Based on NB-IoT and D2D Communications", *ITM Web of Conferences*, vol. 52, pp. 01010–01018, 2023, https://doi.org/10.1051/itmconf/20235201010.

55. Soori, Mohsen, Arezoo, Behrooz and Dastres, Roza. "Internet of Things for Smart Factories in Industry 4.0, A Review", *Internet of Things and Cyber-Physical Systems*, vol. 3, pp. 192–204, 2023, https://doi.org/10.1016/j.iotcps.2023.04.006.

4 Machine Learning in Internet of Things

Sujni Paul and Grasha Jacob

4.1 INTRODUCTION

A new era of technological innovation has begun with the confluence of the Internet of Things (IoT) and Machine Learning (ML), which has completely changed the ways to gather, process, and use data from linked devices. A thorough investigation of this synergistic link is undertaken in this book chapter, revealing its significant influence in a variety of fields. Fundamentally, the chapter seeks to clarify the complex interactions between IoT and ML, providing readers with information and understanding to help them navigate and take advantage of this revolutionary combination. To ensure a strong foundation upon which to develop, a deep knowledge of the fundamental ideas behind both ML and IoT is established.

The potential and problems that come along with the data of IoT, such as data variety, velocity, volume, and accuracy are evaluated next. Additionally, ML techniques— supervised, unsupervised, and reinforcement learning designed specifically for IoT are explored to find how they may be used in various IoT scenarios. A range of fascinating case studies and real-world applications will highlight the practical implications of this integration, showing how ML, among other revolutionary use cases, optimizes energy consumption, improves predictive maintenance procedures, and spurs innovation in the healthcare industry. To enable readers to effectively utilize both technologies in their domains, this chapter aims to give a complete understanding of the dynamic interplay between IoT and ML. It does this by providing information and practical insights. The structure of the chapter is organized as follows.

4.2 FUNDAMENTALS OF IOT AND ML

An outline of the basic concepts of IoT and ML to establish a foundation for understanding their convergence is provided in this section. It explains the architecture and components of IoT systems, including sensors, actuators, gateways, and cloud platforms. The various techniques of ML with their capabilities and applications are also introduced.

DOI: 10.1201/9781032623276-4

4.2.1 Overview of IoT Architecture and Components

IoT architecture is a complete framework built to orchestrate the integration and functionality of varied components in the IoT ecosystem. At its foundation lies the Perception Layer with sensors and actuators that capture real-world data. The Network Layer ensures seamless connectivity through protocols like Wi-Fi and cellular networks, while the Middleware Layer manages data processing, transformation, and device communication. The processed data is then utilized in the Application Layer, which houses specific use case applications that run from smart homes to industrial automation. The Business Layer integrates IoT applications with broader enterprise systems, aligning IoT data with overarching business objectives. Security and Privacy Layer implements crucial measures such as authentication and encryption, safeguarding against unauthorized access. Cloud and Edge Computing Layer involves storing and processing data either in the cloud or closer to the data source. The User Interface Layer provides interfaces and dashboards for end-users to interact with IoT systems. This intricate design enables the creation of scalable, efficient, and secure IoT solutions tailored to diverse applications and industries [1].

4.2.2 Introduction to ML Techniques

ML is a transformative subdomain within the domain of Artificial Intelligence (AI) that empowers computer systems to recognize patterns for making predictions or decisions without being explicitly programmed. ML enables computers to acquire knowledge from data and experiences, thereby improving their performance over time [2]. This transition from traditional rule-based programming to learning from data has opened up new potential across various domains.

One fundamental category of ML is supervised learning. In the case of supervised learning, models have to be trained with labeled datasets, and the test data(input) is then mapped onto the corresponding output labels by the algorithm. For instance, in a supervised learning task of image recognition, the algorithm is trained on a dataset of images with corresponding labels, learning to associate particular features on the images with the correct labels.

Unsupervised learning deals with unlabeled data, looking for patterns or structures within the dataset. Clustering is an unsupervised learning technique in which similar data points are grouped together. An example of unsupervised learning is customer segmentation, where the algorithm identifies groups of customers with analogous purchasing activities without predefined categories.

Reinforcement learning is a technique that is focused on training agents. The agent gets feedback in the form of penalties or rewards, thereby learning the best strategies based on trial and error. This approach is particularly effective in scenarios such as autonomous systems, robotics, and game playing. ML algorithms are classified into several types—neural networks, decision trees, support vector machines (SVM), and ensemble methods. Neural networks consist of nodes that are interconnected to process information in layers. Decision trees are tree-like structures that make decisions based on input features. SVMs are applied for classification and regression tasks, finding the ideal hyperplane to separate different classes. Ensemble methods, like random forests, combine multiple models to enhance overall performance. The success of ML relies heavily on data quality, feature selection, and model evaluation. Additionally, ethical considerations, interpretability, and fairness in ML models have become more and more important. As ML advances, it revolutionizes automated complex tasks, industries, and contributes to solving challenging problems in diverse fields [3].

4.2.3　CONVERGENCE OF IoT AND ML

The convergence of the IoT and ML represents a powerful synergy that amplifies the capabilities of both technologies. At its core, IoT involves the interconnection of physical devices, sensors, and actuators embedded in everyday objects, enabling them to gather and swap data. This vast network generates a massive volume of data that, when properly harnessed, becomes an important resource for ML algorithms.

An important technical aspect of this convergence lies in the data-driven nature of both IoT and ML. IoT devices continuously generate real-time data streams, providing a rich source of information about the environment, user behavior, and system performance. ML algorithms can influence this data to uncover patterns, correlations, and trends that may be challenging for traditional rule-based programming to discern. For instance, prognostic maintenance in industrial IoT can benefit from ML algorithms that analyze sensor data to forecast equipment failures, optimize maintenance schedules, and minimize downtime.

The incorporation of ML models within IoT systems introduces a layer of intelligence that enables devices to adapt, learn, and make autonomous informed decisions. Edge computing plays a vital role in this convergence by bringing ML capabilities closer to the data source, reducing bandwidth requirements and latency. Deploying lightweight ML models directly on IoT devices or edge gateways allows for immediate response and real-time analysis, which is critical in applications like smart cities, healthcare monitoring, and autonomous vehicles.

The merging of IoT and ML introduces challenges related to data privacy, energy efficiency, and security. As sensible information is collected by IoT, ML models need to be designed with privacy-preserving techniques, ensuring that personal data is adequately protected. Security measures, such as encryption and secure communication protocols, become necessary to safeguard both the IoT infrastructure and the ML algorithms.

Hence, the technical convergence of IoT and ML is transforming how we collect, process, and derive insights from data. This integration empowers IoT systems to evolve beyond simple data collection and act as intelligent entities capable of learning from their surroundings. As both technologies continue to advance, their synergy holds enormous prospects in creating novel solutions to diverse domains, from industrial automation to smart homes and beyond [4].

4.3　CHALLENGES AND OPPORTUNITIES IN IOT DATA

The abundance of data that is generated by IoT devices presents both challenges and opportunities. This section discusses the exceptional characteristics of IoT data, including scalability, heterogeneity, and real-time processing requirements, along with the challenges related to handling large volumes of data from diverse sources and emphasizes the potential of ML to haul out valuable insights and patterns from this data.

4.3.1 Data Heterogeneity and Scalability in IoT

One of the primary challenges in managing IoT data lies in its inherent heterogeneity. IoT devices generate diverse types of data, ranging from text and numerical data to image and video feeds. Additionally, these devices often operate on different communication protocols and standards, making it challenging to integrate and analyze data cohesively. This heterogeneity poses obstacles to ensuring interoperability and seamless communication across the IoT ecosystem. Moreover, when the quantity of devices connected increases, the volume of data spawned in excess necessitates robust scalability solutions. Conventional data processing systems find it difficult to work with the increasing influx of data streams, requiring the progress of scalable and efficient infrastructure capable of accommodating the evolving demands of the IoT landscape.

4.3.2 Real-Time Processing Requirements

IoT applications frequently demand real-time processing capabilities to mine appropriate insights and support instantaneous decision-making. In scenarios such as predictive maintenance or autonomous vehicles, delays in data processing can have significant consequences. Meeting these real-time processing requirements involves addressing latency challenges in data transmission, storage, and analytics. Inefficient edge computing solutions allow data to be processed closer to the source instead of relying exclusively on the centralized cloud servers, thereby helping to mitigate latency concerns. This methodology reduces the time taken for data to travel and alleviates network congestion, ensuring that critical decisions can be made promptly based on the most current information available.

4.3.3 The Prospective of IoT Data for ML Applications

While IoT data presents challenges, it brings plenty of opportunities for ML applications. The massive data produced by IoT devices serves as a valuable resource for training and improving ML models. These models can be employed for anomaly detection, predictive analytics, and pattern recognition, enhancing the overall functionality of IoT systems. For instance, in smart homes, ML algorithms can observe user behavior patterns to optimize energy consumption. However, leveraging this potential requires addressing data quality issues, ensuring data privacy and security, and developing ML models capable of handling the dynamic and evolving nature of IoT data. The incorporation of ML with IoT enhances automation and decision-making capabilities and unfold avenues for innovation across various industries.

In navigating the prospects and challenges of IoT data, it is vital for technologists and businesses to adopt a holistic approach, considering factors like data standardization, real-time processing capabilities, scalability, and the strategic integration of ML applications. These aspects, when tackled, expose the full prospects of IoT, paving the way to responsive and more intelligent systems.

4.4 ML TECHNIQUES FOR IOT

4.4.1 SUPERVISED LEARNING IN IoT

In IoT, supervised learning entails using labeled data—data for which the input-output correlations are known—to train ML models. Using the given training dataset, the algorithm in this paradigm learns to make predictions or judgments by mapping input features to corresponding output labels. Thanks to this training process, the model can generalize and make precise predictions on new, untested data. Supervised learning is crucial for IoT applications like predictive maintenance, device behavior classification, and anomaly detection. IoT applications can obtain insights into operational patterns, security concerns, and optimal resource utilization by utilizing algorithms like decision trees, regression models, and SVM.

Supervised learning has an extensive range of uses in IoT, from boosting cybersecurity to increasing industrial process efficiency. For instance, supervised learning algorithms identify user behaviors in IoT-based smart homes to differentiate between benign and possibly dangerous activities. In industrial environments, sensor data is analyzed using predictive maintenance models created by supervised learning to anticipate equipment breakdowns, cutting downtime, and maintenance expenses. Supervised learning approaches enable IoT systems to make well-informed decisions, hence enhancing their overall functionality and reliability in multiple domains.

4.4.1.1 Classification and Regression Algorithms

- **Decision trees (DT)**

DT are an indispensable tool for categorizing IoT devices depending on their behavior. DT can be utilized in the perspective of IoT security to distinguish between legitimate and malicious activity. For example, a decision tree model can detect anomalies and possible security risks by examining the communication and behavior patterns of devices. DTs have the benefit of being simply interpretable making it easier for stakeholders to comprehend the decision-making process [5].

- **SVM**

SVM is an effective technique for categorizing IoT data, particularly in situations involving IoT network intrusion detection. Through input data mapping into a high-dimensional space, SVM determine the optimal hyperplane for classifying data. SVM can be used in an IoT security setting to differentiate between typical and unusual network data. Because of this feature, it's a useful tool for securing IoT networks from online attacks [6].

- **Random Forest**

Overfitting in IoT classification issues is a problem that can be effectively addressed by using Random Forest, an ensemble learning technique. Higher accuracy and resilience are achieved by the model through the combination of numerous decision trees. A Random Forest model can categorize sensor data produced by numerous IoT devices in an application scenario, for example, predicting equipment failures or identifying environmental conditions. Random Forest is an effective tool for processing varied datasets that are typical in IoT scenarios because of its scalability and versatility [7].

4.4.1.2   Use Cases and Applications

- **Decision trees**

By anticipating that equipment is likely to break, predictive maintenance, a crucial IoT application, ensures peak performance and reduces downtime. Models that evaluate sensor data and predict maintenance requirements can be developed using decision trees. Now, let's look at an example to show how a decision tree might be used in an IoT context for predictive maintenance. Figure 4.1 shows the diagrammatical representation of the decision tree.

The decision tree classifier is trained on simulated IoT sensor data, where 'Sensor1' and 'Sensor2' represent sensor readings, and 'MaintenanceNeeded' is a binary label indicating whether maintenance is required. The decision tree is then visualized, allowing interpretation of the decision-making process. This model can be extended to any real-world IoT dataset for predictive maintenance in several industries like manufacturing or healthcare.

- **SVM—anomaly detection in network traffic**

A useful application of SVM is finding anomalies in network traffic in IoT. SVMs are used to differentiate normal and abnormal behavior in network communications because they are good at finding patterns in data and categorizing them. The chart

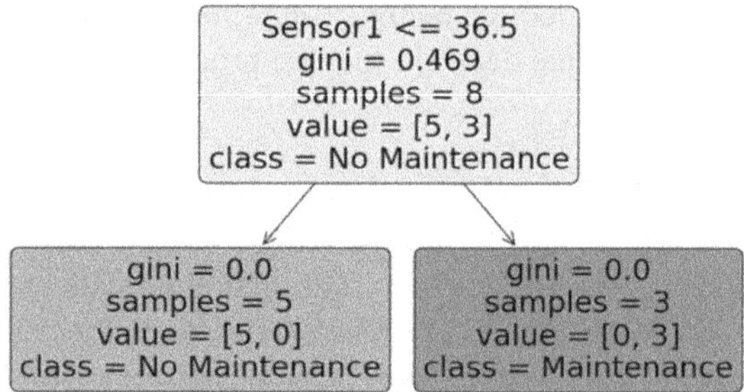

FIGURE 4.1   Decision tree.

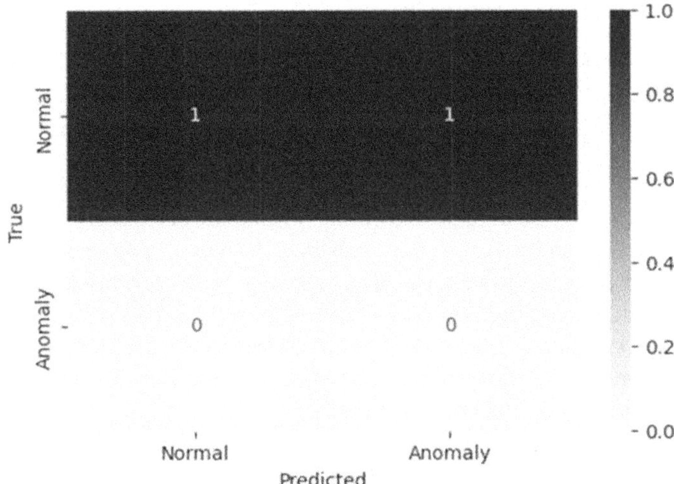

FIGURE 4.2 Confusion matrix.

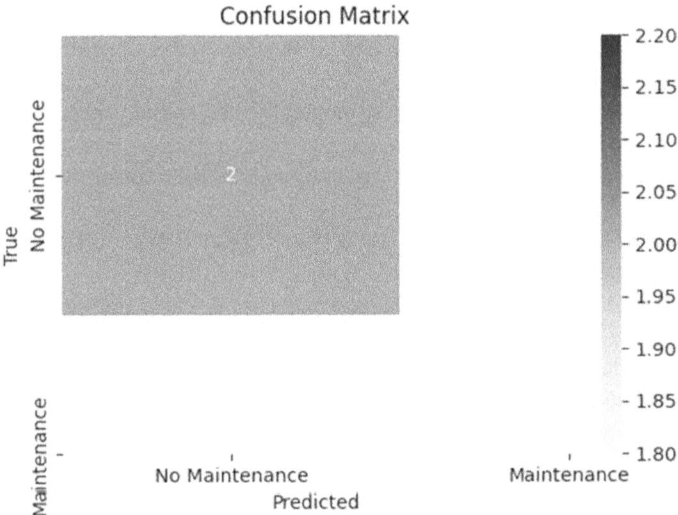

FIGURE 4.3 Predicted value.

given below illustrates anomaly identification by SVM for an IoT network. Figure 4.2 shows the confusion matrix for SVM.

In this example, the SVM classifier is trained on a simulated dataset where different features represent network traffic characteristics, and 'Anomaly' is a binary label indicating whether the traffic is normal or an anomaly. The SVM model is trained to separate normal from anomalous traffic patterns. The confusion matrix visualization provides insights into the model's performance in detecting anomalies. Figure 4.3 shows the predicted value of SVM.

4.4.2 Unsupervised Learning in IoT

Through the usage of more intricate techniques like hierarchical, density-based clustering, and more conventional methods like K-means clustering, readers will get a deeper comprehension of how these algorithms cluster comparable data points and expose underlying patterns and structures. The conversation also covers anomaly detection algorithms, demystifying methods like One-Class SVMs, and Isolation Forests. For identifying anomalies in large and complex data sets produced by IoT devices and to take preventative action against any problems, detection of anomaly is essential.

Here, the strategies for unsupervised learning, with a particular emphasis on anomaly detection and grouping methods are emphasized. The investigation provides readers with a systematic understanding of the revolutionary potential of unsupervised learning in the IoT ecosystem by encompassing theoretical underpinnings, real-world applications, and illustrative use cases.

4.4.2.1 Clustering and Anomaly Detection Algorithms

K-Means clustering algorithm is used to partition a dataset into K distinct, non-overlapping subsets (clusters). This algorithm iteratively assigns each data point to one of the K clusters depending on similar features. The objective is minimizing the within-cluster sum of squares, meaning that the data points contained by each cluster should be as close to each other as possible. The steps of the algorithm are as follows:

- Initialization: Randomly or strategically, K initial cluster centroids are selected.
- Assignment: Each data point is assigned to the cluster whose centroid is closest.
- Update: The centroids of the clusters are updated depending on the assigned data points.
- Repeat: Assignment is iterated, and steps are updated until convergence.

K-Means is scalable and efficient, making it suitable for large datasets. The initial choice of centroids is sensible and may converge to local minima.

4.4.3 Reinforcement Learning in IoT

From the perspective of IoT, reinforcement learning (RL) is a cutting-edge method for enabling adaptive behavior and autonomous decision-making in networked devices. In reinforcement learning (RL), an agent—typically an IoT device or system—learns by interacting with its surroundings to make the best decisions. These choices frequently have to do with resource management, energy efficiency, or adaptation to dynamic situations in the framework of IoT. The agent receives feedback based on its happenings in the form of incentives or penalties, which helps it to improve and hone its decision-making techniques over time. RL has an extensive range of uses in IoT, from resource allocation and traffic optimization in industrial settings to energy management and predictive maintenance. When RL algorithms are incorporated into IoT devices, there is potential for.

4.4.3.1 Learning from Interactions and Feedback

In ML, learning from encounters is a basic idea, especially when it comes to reinforcement learning. According to this paradigm, an agent engages with its settings, acts, and gets feedback in the form of incentives/ punishments. Learning a policy—a collection of guidelines or tactics—that optimizes the cumulative reward over time is the agent's goal. Exploration and exploitation are key components of the learning process, as the agent investigates various actions to determine their effects and uses the knowledge at hand to make the best choices. By means of frequent encounters, the agent enhances its comprehension of the surroundings, modifies its activities, and ultimately elevates its capacity for making decisions.

For many applications, including recommendation algorithms and user interface design, user involvement is essential. Utilizing data produced by user actions to improve system and application performance is known as learning from user interactions. In an online store, for instance, a recommendation engine can deliver tailored recommendations based on consumers' previous interactions with products. With this strategy, systems may adjust to user behavior and personal preferences, resulting in a more personalized and interesting experience. ML algorithms provide insights that enhance user pleasure, boost productivity, and facilitate better decision-making across a range of industries by examining trends in user interactions.

Figure 4.4 chart visually represents how the cumulative rewards augment as the agent is trained from its interactions over episodes.

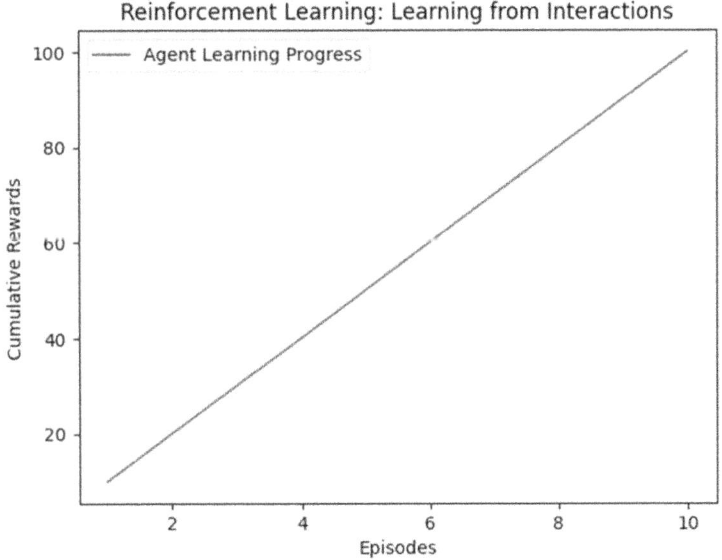

FIGURE 4.4 Cumulative rewards vs episodes.

4.4.3.2 Use Cases and Applications

Energy management in smart buildings is an instance of a reinforcement learning use case within IoT. Imagine an instance where a building is outfitted with an array of IoT devices, including occupancy sensors, lighting controls, and smart thermostats. Optimizing energy consumption and preserving occupant comfort is the goal. In this situation, the RL agent might be a centralized control system that modifies HVAC (Heating, Ventilation, and Air Conditioning) settings in response to outside environmental factors and real-time occupancy trends. Figure 4.5 shows the Reinforcement Learning and Energy Management.

Reinforcement Learning in IoT is applied in Industrial machinery predictive maintenance. Consider a manufacturing plant whose equipment has a network of IoT sensors built into it. Anticipating when equipment repair is required is the aim in order to minimize downtime and lower maintenance expenses. In this case, the RL agent can decide when to plan preventative maintenance by utilizing historical data on machine health, sensor readings, and previous maintenance activities.

In this example, the learning episodes are represented in the x-axis, and the simulated maintenance predictions are represented in the y-axis. When the agent gets trained from historical data interactions, sensor readings, and maintenance records, the chart could demonstrate progress in the accuracy of maintenance predictions over episodes. The goal is to effectively identify patterns indicative of impending machine failures, ultimately optimizing the timing of preventive maintenance actions. Figure 4.6 shows the Reinforcement Learning for Predictive Maintenance.

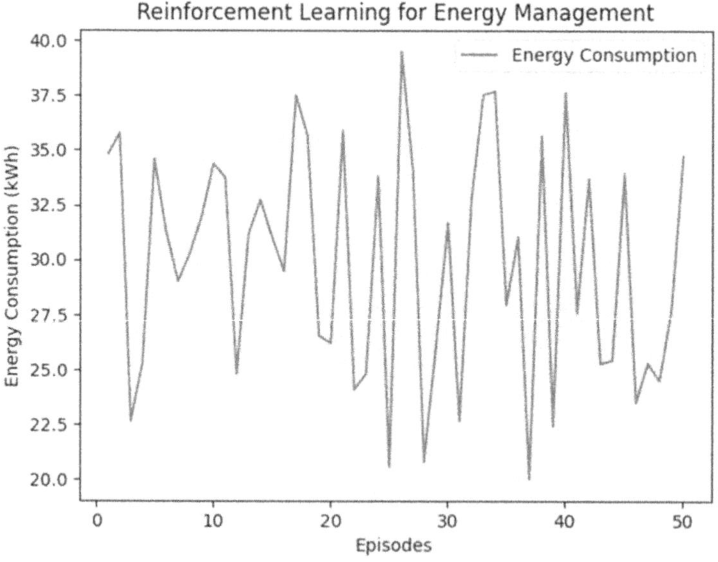

FIGURE 4.5 Reinforcement learning and energy management.

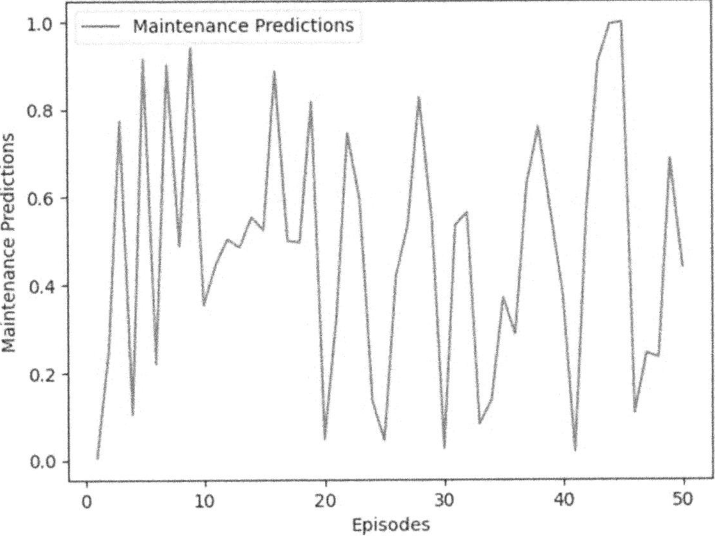

FIGURE 4.6 Reinforcement learning for predictive maintenance.

4.5 CASE STUDIES AND APPLICATIONS

To demonstrate the practical applications of ML in IoT, this section presents multiple case studies and real-world examples. It investigates how ML algorithms can be leveraged for predictive maintenance in industrial IoT, assisting in the active detection of potential breakdowns and optimizing conservation schedules. It further explores the application of ML in anomaly detection in smart cities, facilitating the early detection of unusual events or behaviors. Additionally, the section highlights the role of ML in personalized healthcare monitoring, energy optimization in smart grids, smart transportation, and agriculture and environmental monitoring.

Predictive maintenance in Industrial IoT involves the use of data analytics, ML, and sensor technologies to predict when equipment or machinery is likely to fail, and allowing for timely maintenance to prevent unplanned downtime and optimize maintenance schedules. Let's explore a real-time case study to illustrate this concept.

CASE STUDY 1: PREDICTIVE MAINTENANCE IN A MANUFACTURING PLANT

Scenario:

Imagine a large manufacturing plant that produces automotive components. The plant relies on a variety of machines and equipment to carry out its operations, and any unexpected breakdown can result in significant production losses.

This implementation requires the following

Sensor Deployment:

The considered manufacturing plant is furnished with a network of sensors connected to the IoT platform. These sensors are attached to critical machinery, measuring various parameters like temperature, vibration, pressure, and power consumption.

Collection of Data and Analysis:

The sensors continuously collect data in real time, and this data is transmitted to the IoT platform. Advanced analytics algorithms process the data, identifying patterns and anomalies. ML models are trained on historical data to analyze the normal behavior of the equipment.

Predictive Analytics:

As the ML models are refined and improved over a period of time, they are capable of predicting potential equipment failures. For instance, if abnormal vibration patterns are detected in a certain machine, it may indicate an impending failure.

Alerts and Notifications:

The IoT platform is configured to generate alerts and notifications when a deviation from the normal operating environment is identified. Maintenance teams receive these alerts in real-time and provide insights into which equipment is at risk of failure and the predicted time frame for the failure.

Proactive Maintenance:

Provided with predictive information, maintenance teams can schedule proactive maintenance activities. Instead of relying on fixed schedules or reactive maintenance after a breakdown, they can suggest potential issues before they escalate. This reduces downtime, increases the lifespan of equipment, and minimizes the chances of unexpected failures.

CASE STUDY 2: PERSONALIZED HEALTH MONITORING

In the realm of personalized healthcare monitoring, a real-time case study involves the integration of wearable devices and advanced data analytics to tailor medical insights and interventions to individual patients. Consider a scenario where a patient with chronic conditions, such as diabetes and hypertension, is equipped with a wearable health monitor. This device continuously collects real-time data on vital signs, activity levels, and glucose levels. The data is seamlessly transmitted to a centralized healthcare platform utilizing IoT technology. Advanced analytics algorithms on the platform process this personalized health data, generating dynamic insights into the patient's health status. In the event of deviations from baseline parameters, the system triggers automated alerts for healthcare providers and the patient, facilitating prompt intervention. This proactive approach allows for timely medical adjustments and empowers patients to actively manage their health, fostering a personalized and preventative healthcare model.

Through personalized healthcare monitoring, patients benefit from individualized treatment plans based on real-time data, ultimately improving health outcomes and reducing the risk of complications. This case study exemplifies the transformative potential of IoT-enabled personalized healthcare, illustrating how the convergence of wearable devices and data analytics can usher in a new era of patient-centric and data-driven healthcare management.

4.6 FUTURE DIRECTIONS AND RESEARCH TRENDS

Looking ahead, this section explores the future and research tendencies in ML for IoT. It discusses the concept of federated learning, where models are trained by several IoT devices without centralized data storage, ensuring privacy and efficiency. Additionally, it examines the role of edge computing in enabling decentralized intelligence and reducing the dependence on cloud resources. The section focuses on the importance of explainable AI in IoT, where ML models provide transparent and interpretable outputs. Furthermore, it emphasizes the need for human-machine collaboration and interdisciplinary collaborations to manage the technical and societal challenges posed by ML in IoT.

4.7 CONCLUSION

The chapter accomplishes by reviewing the key points discussed in this chapter. It highlights the transformative impact of ML in enhancing intelligence and efficiency within IoT systems.

REFERENCES

[1] Shi, W., Cao, J., Zhang, Q., Li, Y., & Xu, L. (2011). Edge computing: Vision and challenges. *IEEE Internet of Things Journal*, 3(5), 637–646.

[2] Hastie, T., Tibshirani, R., & Friedman, J. (2009). *The Elements of Statistical Learning: Data Mining, Inference, and Prediction*. Springer.

[3] Russell, S. J., & Norvig, P. (2009). *Artificial Intelligence: A Modern Approach* (3rd ed.). Pearson.

[4] Gubbi, J., Buyya, R., Marusic, S., & Palaniswami, M. (2013). Internet of Things (IoT): A vision, architectural elements, and future directions. *Future Generation Computer Systems*, 29(7), 1645–1660.

[5] Breiman, L. (1984). *Classification and Regression Trees*. CRC Press.

[6] Cortes, C., & Vapnik, V. (1995). Support-vector networks. *ML*, 20(3), 273–297.

[7] Liaw, A., & Wiener, M. (2002). Classification and regression by random forest. *R News*, 2(3), 18–22.

5 Role of Machine Learning in Real-Life Environment

*Balasaraswathi V.R., Nathezhtha T.,
Vaissnavie V., and Rajashree*

5.1 INTRODUCTION

Machine learning (ML) has a crucial role in leveraging technologies around Artificial Intelligence (AI). ML is frequently referred to as AI due to its capacity for learning and decision-making, but in truth, it is a subset of AI. Image Processing and Natural language processing (NLP) are examples of machine learning approaches that can be used to speed up data collection and format data. This can improve the ability to spot clinical patterns and help with more accurate forecasts. Pattern recognition and data analysis are a sort of data mining (DM) that enables computers to "learn" on their own. Numerous advantages of ML include enhanced decision-making, accuracy, and effectiveness. ML has emerged as a powerful tool in many fields, growing businesses, and a wide range of applications [1]. Following are the categories of machine learning.

Supervised Learning: In supervised learning, the input is coupled with target labels, and the system learns from labelled samples. The objective is to train the model to use the labels provided to transfer the input data to the desired output. In order to forecast or categorise previously unobserved data, the algorithm generalises from the labelled data. Supervised learning algorithms encompass a variety of methods, such as Linear Regression (LR), Logistic Regression, Decision Trees (DT), and SVM [2].

Unsupervised Learning: It is the process of finding patterns, structures, or relationships in unlabelled data without using explicit target labels. The objective is to gather relevant data points, minimise the complexity of the data, or draw insightful conclusions. It is used for anomaly detection, dimensionality reduction, and data clustering. Clustering techniques, such as k-means clustering and hierarchical clustering, are employed to group similar data points based on their features. Methods like principal component analysis (PCA) and t-distributed stochastic neighbour embedding (t-SNE) are utilised to reduce the number of features while retaining essential information. Anomaly detection techniques locate out-of-the-ordinary or rare data points [3].

DOI: 10.1201/9781032623276-5

Reinforcement Learning: Through iterative experimentation, an agent acquires the knowledge of interacting with the environment to optimise cumulative rewards. The learning process involves the agent actively engaging with the environment and receiving feedback in the form of rewards or punishments. The ultimate goal is to discern optimal courses of action or policies that yield the highest cumulative reward over time. Reinforcement learning has found successful applications in diverse areas such as robotics, autonomous systems, and gaming. The agent, environment, actions, rewards, and a policy that directs the agent's decision-making process are crucial elements in reinforcement learning [4].

5.2   MACHINE LEARNING IN HEALTHCARE

A variety of disease identification and treatments has been revolutionised by machine learning, which has become a potent instrument in the medical field. ML algorithms can analyse complicated patterns, generate precise predictions, and help healthcare practitioners make educated decisions by utilising large volumes of patient data and advanced algorithms. Healthcare with the help of ML is benefitted in many fields, as shown in Figure 5.1.

5.2.1   USE OF HEALTHCARE WITH ML

5.2.1.1   Improved Diagnosis

Large datasets including electronic health records, genetic data, and medical imaging are used for analysis by machine learning algorithms. These algorithms are able to

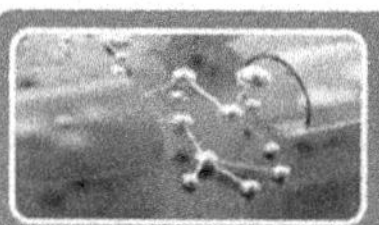

FIGURE 5.1   Role of ML in healthcare.

recognise small patterns that would not be noticeable to human observers, resulting in quicker and more accurate diagnoses. For instance, ML models in radiology can discover anomalies in medical pictures or early indicators of diseases like cancer, allowing for early intervention and better patient outcomes [5].

Liang et al. introduced an innovative classification framework known as Multi-Level FC Fusion (MFC) to identify brain diseases. The model is developed utilising both supervised and unsupervised learning techniques and methodical functional Magnetic Resonance Imaging dataset tests. The effectiveness and generality of the method are shown by the results, demonstrating strong classification performance across various preprocessing pipelines and cross-validation schemes, which is a characteristic feature of the proposed approach [6].

5.2.1.2 Personalised Treatment

Personalised treatment plans are one of machine learning's main advantages in the medical field. ML algorithms can forecast the best treatment options and dosages by considering patient-specific aspects like medical history, genetic information, and lifestyle data. This individualised strategy raises patient happiness, decreases unfavourable reactions, and improves therapeutic effectiveness. In order to implement pre-emptive interventions and preventive care plans, machine learning models can help identify patients who are more likely to experience the onset of particular illnesses [7].

The Internet of Things, wearable technology, genetics, and image processing marked the beginning of the digital era in medicine. AI technologies are required to enable forecasts in order to personalise therapies. Concerns, including explainability, liability, and privacy must be addressed for AI to be widely used in healthcare [8].

5.2.1.3 Predictive Analytics

Based on past data, machine learning algorithms can forecast disease development and patient outcomes. These algorithms may produce precise predictions about disease trajectories and identify high-risk individuals by examining a massive quantity of patient data, including demographics, symptoms, test results, and therapy responses. Predictive analytics can assist healthcare professionals in better resource allocation, early detection of probable issues, and patient management techniques. Ahmed et al. proposed a fuzzy model to diagnose the diabetic level with a membership value [9]. Many researchers focus on efficient prediction of cancer, which helps in the early detection of disease and minimises the death rate [10].

5.2.1.4 Drug Discovery and Development

ML is also used in the field of finding new drugs and developing them [11]. ML algorithms are capable of identifying promising medication candidates, predicting their efficacy, and even optimising dosages by studying huge databases of molecular structures, biological information, and clinical trial outcomes. This method speeds up the research process, lowers expenses, and improves the likelihood that novel treatments will be developed successfully.

Finding numerous novel pharmacological indications has been greatly aided by chance. Drug-repositioning hypotheses could be created and validated with the aid of algorithmic identification of patient-reported unexpected drug use in social media [12].

5.2.1.5 Streamlining Healthcare Operations

Beyond diagnosis and treatment, ML is revolutionising administrative and operational activities in the healthcare industry. The documentation and coding processes can be made simpler by using NLP algorithms to analyse medical data and extract pertinent information. To increase operational effectiveness in healthcare settings, machine learning models can also optimise resource allocation, forecast patient flow, streamline scheduling, and other techniques to detect the patient's stability [13].

By improving patient care overall and patient diagnosis, treatment, and care, machine learning has the potential to completely transform the healthcare industry. Machine learning can help healthcare providers make more accurate diagnoses, provide individualised treatment regimens, forecast disease outcomes, and streamline operations by utilising the power of data and sophisticated algorithms. However, ensuring the responsible and secure application of machine learning in healthcare still depends on ethical considerations, privacy issues, and the requirement for human oversight. As this discipline develops, it holds the potential to guide in a new era of precision medicine and better patient outcomes across the board.

The research articles about machine learning in relation to healthcare are presented in Table 5.1, which describes the diseases identified with its associated dataset and the advantages of the proposed article. One of the widely used datasets for breast cancer identification is Welikala et al. [14]. The Wisconsin Breast Cancer (Diagnosis) dataset comprises 569 instances with 32 attributes, including an ID and a target variable. Similarly, the Wisconsin Breast Cancer (Prognosis) dataset includes 198 instances and 34 attributes, each with an ID and a target variable. Notably, the forecast dataset underwent the removal of four instances with missing attribute values. The forecast dataset exhibits a notable skew, with 151 instances of non-recurring outcomes and 47 instances of recurring outcomes. To address the challenge of imbalanced categorization, two approaches, namely the algorithmic and data-centric approaches, were employed in both the BC Wisconsin Diagnostic and Prognostic datasets. The first tactic involves training the model to improve performance in minority classes by employing cost-sensitive learning or applying a misclassification penalty. The results [15,16] are depicted in Figure 5.2.

5.3 MACHINE LEARNING IN BUSINESS

Machine learning has spread across many facets of business, giving organisations useful insights, automation capabilities, and enhanced decision-making procedures. Some crucial business applications of ML are shown in Figure 5.3. Business processes are assets of the organisation that are essential for providing value to the consumers [20], and evolving technologies like ML and Deep Learning(DL) play a vital role in the elevation of business.

TABLE 5.1
Survey of ML in Healthcare

References	Disease	ML Techniques	Datasets	Conclusion	Advantages
[15]	Cancer	Logistic Regression, KNN, SVM, RF, DT, and Naïve Bayes (NB) classification	Wisconsin Breast Cancer Dataset	Importance early diagnoses experienced by both current and former breast cancer patients.	Early diagnosis can reduce the risk of death
[16]	Breast, lung, and cervical cancer Datasets	Genetic Algorithm-Correlation Based Feature Selection. DT, SVM, Linear Discriminant Analysis (LDA), NN	Irvine repository	Provides physicians with a diagnostic tool by helping them to make a proper diagnosis	It acts as a general tool for extracting patterns from several clinical trials for various cancer diseases.
[17]	Oral Potentially Malignant Disorders (OPMD)	Random Forest	Surgical Care Unit (SCU) Dataset	Predict the risk of cancer for Oral Potentially Malignant Disorders. Sensitivity: 0.82 Specificity: 0.9	It is a unique model for accurate and affordable OPMDs/ Oral Squamous Cell Carcinoma (OSCC) screening. Cost-effective tool for predicting OPMDs.
[18]	Oral Cavity Squamous-Cell Carcinoma (OCSCC)	Naïve Bayes NB, Bagging of Naive Bayes, KNN, J48, boosting J48	Patients of University of Naples	92 % Tumour level and the prediction of nodal status in patients with oral cavity squamous cell carcinoma (OCSCC) and oropharyngeal cancers is anticipated.	Accurate feature subset is identified for predicting the cancer in an accurate manner
[19]	Coronavirus disease	SVM	Epidemiology Dataset	87% accurately categorise the patient's condition into no infection, mild infection, and serious infection based on the infection level.	Accurate prediction

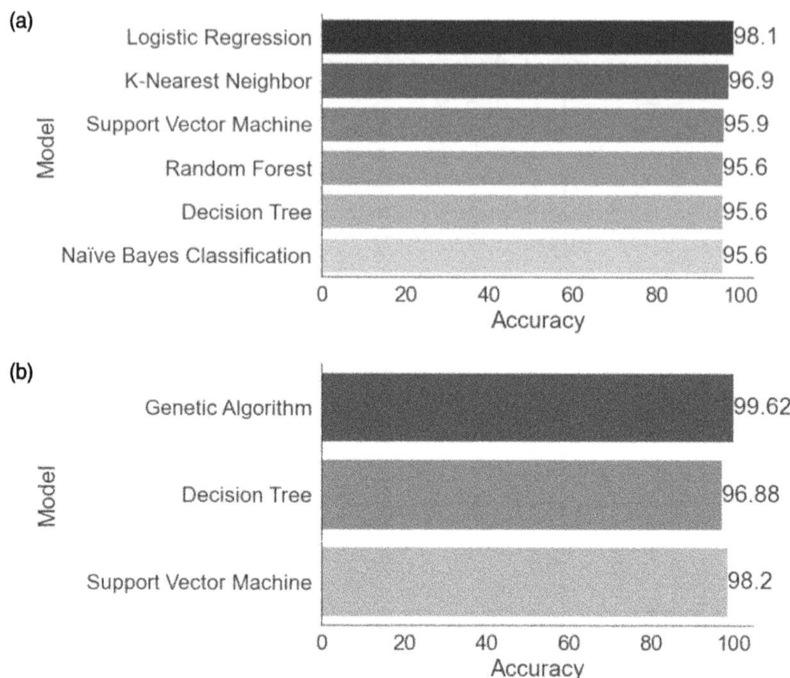

FIGURE 5.2 (a) Accuracy of breast cancer prediction for Wisconsin dataset (b) accuracy of cancer prediction Irvine dataset.

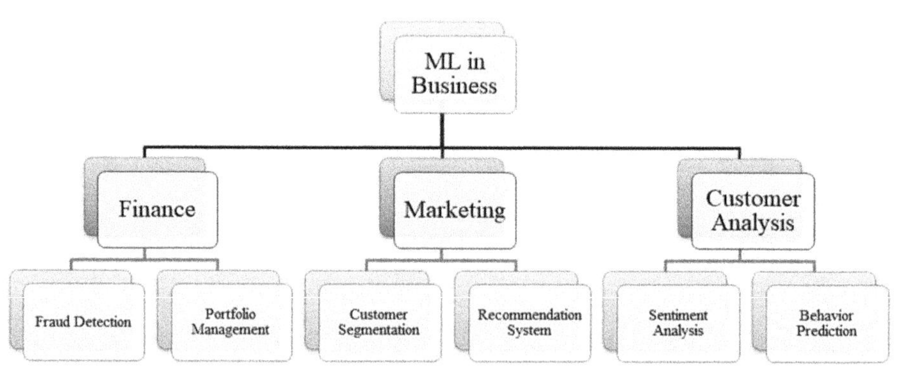

FIGURE 5.3 Role of machine learning in business.

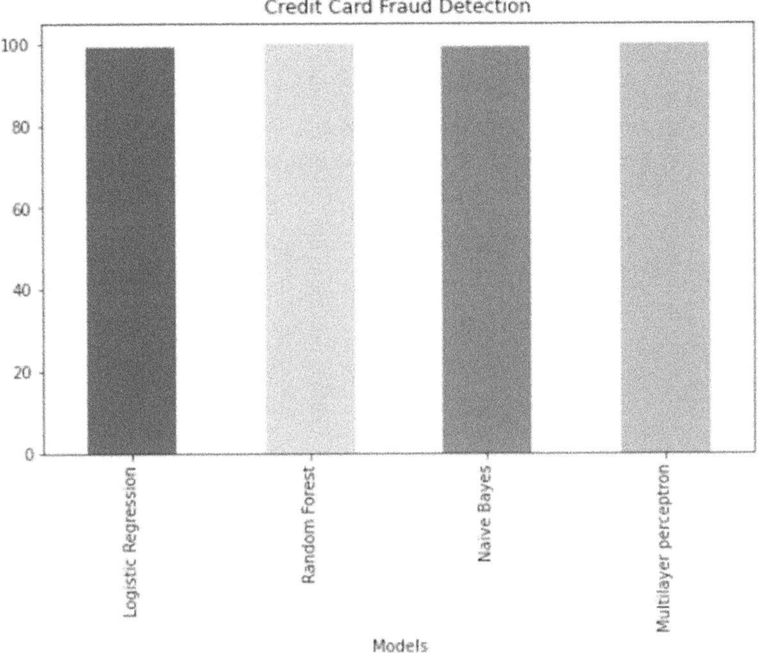

FIGURE 5.4 Accuracy of credit card fraud detection compared with the results.

Credit Card Fraud Detection

For computational intelligence algorithms, detecting fraud in credit card transactions is a difficult problem that involves issues including idea drift, class imbalance, and verification latency. Most learning algorithms include assumptions that are false when used with fraud-detection systems (FDS) in the actual world. The fraud detection challenge is formalised in this work by outlining the operational parameters of FDSs that process enormous volumes of credit card transactions [21].

For both cardholders and financial institutions, credit card fraud causes considerable financial loss. This study tries to identify high-class imbalanced data and changes in the nature of fraud. Figure 5.4 shows the accuracy of credit card fraud detection compared with the results. The European Card Benchmark dataset was used for comparative analysis, with ML methods to increase accuracy [22].

Healthcare systems are at risk from collusive fraud, which occurs when several fraudsters work together to steal money from health insurance funds. Due to the similarities to medical visits and the lack of labelled data, current approaches are facing difficulty in detecting fraud. To identify suspicious patient behaviour, Fraud Auditor, a three-stage visual analytics approach, integrates interactive co-visit networks, enhanced community detection algorithms, and custom visualisations. The method's efficacy and usability were demonstrated through real-world examples and professional commentary [23]. Table 5.2 contains the widely used application for business and the techniques used with its dataset.

TABLE 5.2
Application for Business and the Techniques Used with Its Dataset

References	Application	ML Techniques	Datasets	Results	Conclusion	Advantages
[24]	Credit Card Fraud Detection	LR, RF, NB and Multilayer Perceptron (MLP)	Credit Card Fraud Detection	LR: 99.23% RF:99.96% NB:99.23% MLP:99.93%	Solves business problem.	Helps effectively to detect and prevent frauds
[25]	Fake Image Detection	KNN, DT, Gaussian Naïve Bayes, RF, Adaboost Classifier, and LR	In-house dataset	Accuracy: 99%	RF is comparatively better	Comparative results provide a framework
[26]	Fraud Detection	Hybrid Technique- Combination of Supervised and Un-Supervised learning	Credit Card Fraud Detection Dataset.	Evaluated from baseline to 5000 clusters	Outliers highlights the information about risk	Better results
[27]	Employee Churn prediction and Retention	CatBoost, RF, LR, DT, SVM	Human Resource Information Dataset (HRIS)	Accuracy:95.4%	CatBoost Algorithm gives higher prediction rate	Predicts the risk of churn in any organisation
[28]	Recommendation systems	NB, SVM	Movie Lens Dataset	NA	NB performs better	Performs sentiment analysis

5.4 MACHINE LEARNING IN ROBOTICS AND AUTONOMOUS SYSTEMS

Robotics and autonomous systems depend on machine learning to perceive, acquire, and adapt to their surroundings. Robots can handle challenging situations and make intelligent decisions using sensors and data analysis. Robots can benefit from machine learning algorithms for tasks like object detection, path planning, and obstacle avoidance. Robots can continuously enhance their performance and refine their actions by learning from real-time data. ML is a key component of autonomous vehicles' ability to sense their environment, decipher traffic signs, and make quick decisions. With the help of this technology, self-driving cars can manoeuvre safely and effectively, lowering the risk of accidents [29–31]. Table 5.3 shows the survey of ML in robotics and autonomous systems, Figure 5.5 (a) specifies the accuracy of robotic-assisted urologic surgery, Figure 5.5 (b) shows the pedestrian detection in an autonomous vehicle environment, Figure 5.5 (c) provides the analysis of crash severity, and Figure 5.5 (d) shows the citrus fruit disease detection.

Machine learning algorithms are used to recognise and track specific objects or features in the robot's field of view [37]. This capability is essential for tasks such as picking and placing objects, manipulating, or following a moving target. ML can be used to optimise motion planning and control algorithms [38]. Reinforcement learning (RL) algorithms can learn optimal control policies by trial and error to perform tasks with more efficiency and adaptability. This is particularly useful in applications like robot arm manipulation or mobile robot navigation. ML enables robots to interact with humans in natural and intuitive ways. NLP techniques can be used to understand and respond to human speech, while sentiment analysis can help robots gauge user emotions. This enhances the robot's ability to collaborate and communicate effectively with humans. Machine learning algorithms are instrumental in autonomous navigation for robots. By analysing sensor data and learning from past experiences, it can navigate complex environments, avoid obstacles, and plan efficient paths. This is crucial for applications such as autonomous drones, self-driving cars, or warehouse robots. Figure 5.6 shows the different applications of ML in Robotics.

Machine learning techniques can optimise task planning and scheduling for robots. By analysing historical data, robots can learn to estimate task durations, allocate resources efficiently, and improve overall task execution. Machine learning models can analyse environmental data, including soil moisture levels, weather conditions, and crop requirements, to optimise irrigation scheduling. By learning from historical data, these algorithms can make predictions and adapt irrigation practices to conserve water and improve crop health. The survey of ML in NLP is shown in Table 5.4.

NLP enables computers to comprehend, examine, and produce human language and depends heavily on machine learning. Using ML approaches for sentiment analysis, text data can be classified as having positive, negative, or neutral feelings. By training on labelled data, these algorithms can accurately determine the sentiment expressed in customer reviews, social media posts, or other text sources. Businesses can utilise this to gain deeper insight into the attitudes and viewpoints of a sizable client base.

TABLE 5.3
Survey of ML in Robotics and Autonomous Systems

References	Application	ML Techniques	Datasets	Results	Conclusion	Advantages
[32]	Robotic-assisted urologic surgery	Combination of ML and Augmented Reality (AR)	Johns Hopkins University - Intelligent Systems Institute (JHU-ISI) Gesture and Skill Assessment Working Set (JIGSAWS)	The accuracy, sensitivity, and specificity metrics achieve rates of 92.5%, 95.8%, and 88.8%, respectively.	ML-guided autonomous surgery	Surgeon deals the novel data
[33]	Pedestrians detection in autonomous vehicle environment	KNN, SVM and NB Classifier	Sample points generated by 3D Light Detection and Ranging (LIDAR)	Accuracy KNN:90.72% SVM Linear: 96.29 SVM Quadratic: 89.28%	Provides more accurate results at all type lighting conditions	Scalability of ML enables the detection of dissimilar objects
[34]	Analyses the crash severity	Classification and regression tree (CART), SVM, eXtreme Gradient Boosting (XGBoost)	Crash reports received in California from Jan- 2019 to Oct- 2020	Accuracy XGBoost:64.10% CART:66.67 % SVM:71.79 %	XGBoost model recognises the injured crashes in an efficient manner	Reduces the severity of AV-involved crashes
[35]	Detects the citrus fruit disease automatically	K Means clustering, Artificial Neural Network(ANN) and SVM	Citrus fruit dataset	Accuracy ANN:88.96% SVM: 93.12%	Detects the citrus infection in less computational effort	Powerful tool for automated detection
[36]	Automation of Farm Irrigation System	SVM and Support Vector Regression(SVR)	Agriculture Crop Production in India Dataset	–	Economical approach for automation in agriculture	Computationally efficient approach

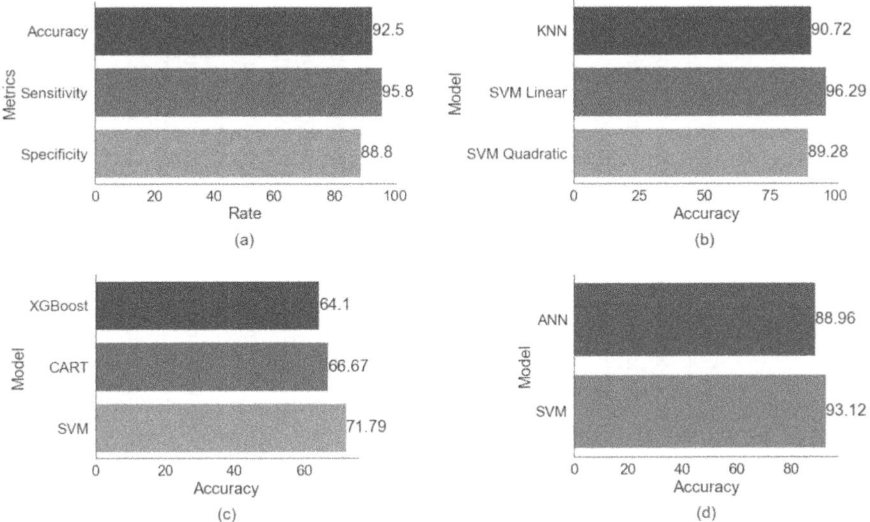

FIGURE 5.5 Accuracy of (a) robotic assisted urologic surgery (b) pedestrians detection in autonomous vehicle environment (c) analysis of crash severity (d) citrus fruit disease detection.

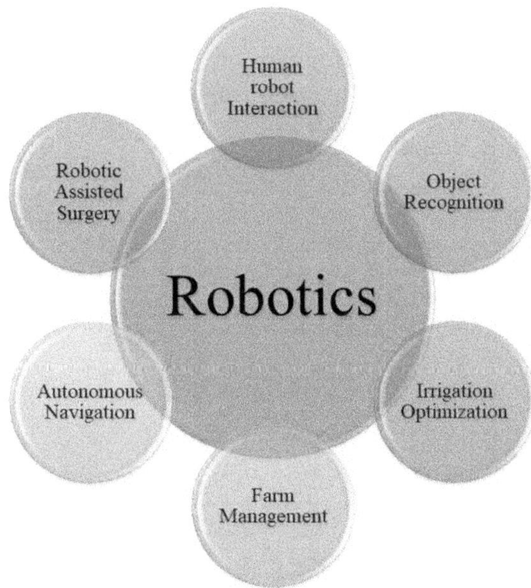

FIGURE 5.6 Applications of ML in robotics.

TABLE 5.4
Survey of ML in NLP

References	Application	ML Techniques	Datasets	Results	Conclusion	Advantages
[39]	Sentiment analysis using online reviews	Unsupervised learning	Airport and Airline reviews, tourist reviews, etc.	–	Analyses the online reviews in the domain of tourism and hospitality.	Helps to develop the managerial strategies for consumers.
[40]	Determines the users view in an automated way	Binary Classification	Social media platform like twitter and reddit	Twitter Positive Opinion:61.66% Negative Opinion:38.34% Reddit Positive opinion: 71.72% Negative opinion: 28.28%	ML models identify the sentiments in the data and also determine the level of acceptance.	Negative opinion helps to identify the key fears and reservation.
[41]	Automatically symbolises the expression of feelings for movie,	DT, Bernoulli NB, SVM, Maximum Entropy (ME), and Multinomial NB.	Movie Review Dataset	Accuracy BNB: % DT: 80.17% SVM:87.33% ME: 60.67% MNB: 88.50%	MNB provides better result	Helps to realise the expression of feelings around everything like product, social media, etc.
[42]	Analyse and categorises the Arabic text	MNB, Bernoulli NB, Stochastic Gradient Descent (SGD), LR, SVM classifier	Al-Khaleej dataset	SVM performs better	Helps to maintain the significant information in several domains.	Useful for multiple domains, mostly on social media.

Machine learning models can find and classify named entities within a text, such as people, organisations, places, and dates. Named Entity Recognition (NER) can be useful for applications, including information extraction, search engines, and recommendation systems. Machine learning algorithms are trained to accurately recognise and extract named things from annotated data.

Text documents are categorised into predetermined groups by machine learning algorithms for purposes including spam detection, subject categorisation, sentiment analysis, and content filtering.

Machine translation systems improve with machine learning approaches like neural machine translation (NMT), enhancing accuracy by recognising contextual cues and grammatical patterns in multiple languages.

Question-answering systems use machine learning techniques to analyse and provide answers to questions using text matching, natural language comprehension, and information retrieval, enabling them to understand queries and extract relevant information.

REFERENCES

1. S. V. Mahadevkar et al., "A Review on Machine Learning Styles in Computer Vision—Techniques and Future Directions," *IEEE Access*, vol. 10, pp. 107293–107329, 2022, DOI: 10.1109/ACCESS.2022.3209825.
2. J. Camacho, G. Maciá-Fernández, N. M. Fuentes-García and E. Saccenti, "Semi-Supervised Multivariate Statistical Network Monitoring for Learning Security Threats," *IEEE Transactions on Information Forensics and Security*, vol. 14, no. 8, pp. 2179–2189, Aug. 2019, DOI: 10.1109/TIFS.2019.2894358.
3. S. J. Pan and Q. Yang, "A Survey on Transfer Learning," *IEEE Transactions on Knowledge and Data Engineering*, vol. 22, no. 10, pp. 1345–1359, Oct. 2010, DOI: 10.1109/TKDE.2009.191.
4. X. Ma and W. Shi, "AESMOTE: Adversarial Reinforcement Learning with SMOTE for Anomaly Detection," *IEEE Transactions on Network Science and Engineering*, vol. 8, no. 2, pp. 943–956, 1 Apr.–June 2021, DOI: 10.1109/TNSE.2020.3004312.
5. M. Li and Z.-H. Zhou, "Improve Computer-Aided Diagnosis with Machine Learning Techniques Using Undiagnosed Samples," *IEEE Transactions on Systems, Man, and Cybernetics – Part A: Systems and Humans*, vol. 37, no. 6, pp. 1088–1098, Nov. 2007, DOI: 10.1109/TSMCA.2007.904745.
6. Y. Liang and G. Xu, "Multi-Level Functional Connectivity Fusion Classification Framework for Brain Disease Diagnosis," *IEEE Journal of Biomedical and Health Informatics*, vol. 26, no. 6, pp. 2714–2725, June 2022, DOI: 10.1109/JBHI.2022.3159031.
7. K. Paranjape, M. Schinkel and P. Nanayakkara, "Short Keynote Paper: Mainstreaming Personalized Healthcare – Transforming Healthcare through New Era of Artificial Intelligence," *IEEE Journal of Biomedical and Health Informatics*, vol. 24, no. 7, pp. 1860–1863, July 2020, DOI: 10.1109/JBHI.2020.2970807.
8. S. Zhang, S. M. H. Bamakan, Q. Qu and S. Li, "Learning for Personalized Medicine: A Comprehensive Review from a Deep Learning Perspective," *IEEE Reviews in Biomedical Engineering*, vol. 12, pp. 194–208, 2019, DOI: 10.1109/RBME.2018.2864254.

9. U. Ahmed, et al., "Prediction of Diabetes Empowered with Fused Machine Learning," *IEEE Access*, vol. 10, pp. 8529–8538, 2022, DOI: 10.1109/ACCESS.2022.3142097.

10. N. Fatima, L. Liu, S. Hong and H. Ahmed, "Prediction of Breast Cancer, Comparative Review of Machine Learning Techniques, and Their Analysis," *IEEE Access*, vol. 8, pp. 150360–150376, 2020, DOI: 10.1109/ACCESS.2020.3016715.

11. N. R. C. Monteiro, B. Ribeiro and J. P. Arrais, "Drug-Target Interaction Prediction: End-to-End Deep Learning Approach," *IEEE/ACM Transactions on Computational Biology and Bioinformatics*, vol. 18, no. 6, pp. 2364–2374, 1 Nov.–Dec. 2021, DOI: 10.1109/TCBB.2020.2977335.

12. B. Ru, D. Li, Y. Hu and L. Yao, "Serendipity—A Machine-Learning Application for Mining Serendipitous Drug Usage from Social Media," *IEEE Transactions on NanoBioscience*, vol. 18, no. 3, pp. 324–334, July 2019, DOI: 10.1109/TNB.2019.2909094.

13. B.-S. Lin, et al., "Fall Detection System with Artificial Intelligence-Based Edge Computing," *IEEE Access*, vol. 10, pp. 4328–4339, 2022, DOI: 10.1109/ACCESS.2021.3140164.

14. R. A. Welikala, et al., "Automated Detection and Classification of Oral Lesions Using Deep Learning for Early Detection of Oral Cancer," *IEEE Access*, vol. 8, pp. 132677–132693, 2020, DOI: 10.1109/ACCESS.2020.3010180.

15. Ak, Muhammet Fatih, "A Comparative Analysis of Breast Cancer Detection and Diagnosis Using Data Visualization and Machine Learning Applications," *Healthcare*, vol. 8., no. 2, pp. 1–23, 2020.

16. Hsu, Ching-Hsien, et al., "Effective Multiple Cancer Disease Diagnosis Frameworks for Improved Healthcare Using Machine Learning," *Measurement*, vol. 175, pp. 109–145, 2021.

17. Wang, Xiangjian, et al., "A Personalized Computational Model Predicts Cancer Risk Level of Oral Potentially Malignant Disorders and Its Web Application for Promotion of Non-invasive Screening," *Journal of Oral Pathology & Medicine*, vol. 49, no. 5, pp. 417–426, 2020.

18. Romeo, Valeria, et al., "Prediction of Tumor Grade and Nodal Status in Oropharyngeal and Oral Cavity Squamous-Cell Carcinoma Using a Radiomic Approach," *Anticancer Research*, vol. 40, no. 1, pp. 271–280, 2020.

19. Guhathakurata, Soham, et al., "A Novel Approach to Predict COVID-19 Using Support Vector Machine," *Data Science for COVID-19*, vol. 1, pp. 351–364, 2021.

20. G. Polančič, S. Jagečič and K. Kous, "An Empirical Investigation of the Effectiveness of Optical Recognition of Hand-Drawn Business Process Elements by Applying Machine Learning," *IEEE Access*, vol. 8, pp. 206118–206131, 2020, DOI: 10.1109/ACCESS.2020.3034603.

21. A. Dal Pozzolo, G. Boracchi, O. Caelen, C. Alippi and G. Bontempi, "Credit Card Fraud Detection: A Realistic Modeling and a Novel Learning Strategy," *IEEE Transactions on Neural Networks and Learning Systems*, vol. 29, no. 8, pp. 3784–3797, Aug. 2018, DOI: 10.1109/TNNLS.2017.2736643.

22. F. K. Alarfaj, I. Malik, H. U. Khan, N. Almusallam, M. Ramzan and M. Ahmed, "Credit Card Fraud Detection Using State-of-the-Art Machine Learning and Deep Learning Algorithms," *IEEE Access*, vol. 10, pp. 39700–39715, 2022, DOI: 10.1109/ACCESS.2022.3166891.

23. J. Zhou, et al., "FraudAuditor: A Visual Analytics Approach for Collusive Fraud in Health Insurance," *IEEE Transactions on Visualization and Computer Graphics*, vol. 29, no. 6, pp. 2849–2861, 1 June 2023, DOI: 10.1109/TVCG.2023.3261910.

24. Armedja, Dejan, et al., "Credit card fraud detection-machine learning methods," *International symposium* INFOTEH-JAHORINA (INFOTEH), 2019.

25. Hamid, Yasir, et al., "An improvised CNN model for fake image detection," *International Journal of Information Technology*, vol. 15, no. 1, pp. 5–15, 2023.

26. Carcillo, Fabrizio, et al,. "Combining unsupervised and supervised learning in credit card fraud detection," *Information Sciences*, vol. 557, pp. 317–331, 2021.

27. Jain, Nishant, Abhinav Tomar and Prasanta K. Jana, "A Novel Scheme for Employee Churn Problem Using Multi-Attribute Decision Making Approach and Machine Learning," *Journal of Intelligent Information Systems*, vol. 56, pp. 279–302, 2021.

28. Marappan, Raja and S. Bhaskaran, "Movie Recommendation System Modeling Using Machine Learning," *International Journal of Mathematical, Engineering, Biological and Applied Computing*, pp. 12–16, 2022.

29. G. Jia, H.-K. Lam, S. Ma, Z. Yang, Y. Xu and B. Xiao, "Classification of Electromyographic Hand Gesture Signals Using Modified Fuzzy C-Means Clustering and Two-Step Machine Learning Approach," *IEEE Transactions on Neural Systems and Rehabilitation Engineering*, vol. 28, no. 6, pp. 1428–1435, June 2020, DOI: 10.1109/TNSRE.2020.2986884.

30. A. A. Alhussan, D. S. Khafaga, E.-S. M. El-Kenawy, A. Ibrahim, M. M. Eid and A. A. Abdelhamid, "Pothole and Plain Road Classification Using Adaptive Mutation Dipper Throated Optimization and Transfer Learning for Self Driving Cars," *IEEE Access*, vol. 10, pp. 84188–84211, 2022, DOI: 10.1109/ACCESS.2022.3196660.

31. A. Altameem, A. Kumar, R. C. Poonia, S. Kumar and A. K. J. Saudagar, "Early Identification and Detection of Driver Drowsiness by Hybrid Machine Learning," *IEEE Access*, vol. 9, pp. 162805–162819, 2021, DOI: 10.1109/ACCESS.2021.3131601.

32. Ma, Runzhuo, et al., "Machine Learning in the Optimization of Robotics in the Operative Field," *Current Opinion in Urology*, vol. 30, no. 6, p. 808, 2020.

33. Navarro, Pedro J., et al., "A Machine Learning Approach to Pedestrian Detection for Autonomous Vehicles Using High-Definition 3D Range Data," *Sensors*, vol. 17, no. 1, pp. 1–15, 2016.

34. Chen, Hengrui, et al., "Exploring the Mechanism of Crashes with Autonomous Vehicles Using Machine Learning," *Mathematical Problems in Engineering*, vol. 1, pp. 1–10, 2021.

35. Doh, Benjamin, et al., "Automatic Citrus Fruit Disease Detection by Phenotyping Using Machine Learning," *25th International Conference on Automation and Computing (ICAC)*, pp. 1–5, 2019.

36. Vij, Anneketh, et al., "IoT and Machine Learning Approaches for Automation of Farm Irrigation System," *Procedia Computer Science*, vol. 1, pp. 1250–1257, 2020.

37. S. A. Pedram, P. W. Ferguson, M. J. Gerber, C. Shin, J.-P. Hubschman and J. Rosen, "A Novel Tissue Identification Framework in Cataract Surgery Using an Integrated Bioimpedance-Based Probe and Machine Learning Algorithms," *IEEE Transactions on Biomedical Engineering*, vol. 69, no. 2, pp. 910–920, Feb. 2022, DOI: 10.1109/TBME.2021.3109246.

38. D.-P. Tran, G.-N. Nguyen and V.-D. Hoang, "Hyperparameter Optimization for Improving Recognition Efficiency of an Adaptive Learning System," *IEEE Access*, vol. 8, pp. 160569–160580, 2020, DOI: 10.1109/ACCESS.2020.3020930.

39. Jain, Praphula Kumar, Rajendra Pamula and Gautam Srivastava, "A Systematic Literature Review on Machine Learning Applications for Consumer Sentiment Analysis Using Online Reviews," *Computer Science Review*, vol. 41, pp. 100413, 2021.

40. Bakalos, Nikolaos, Nikolaos Papadakis and Antonios Litke, "Public Perception of Autonomous Mobility Using ML-Based Sentiment Analysis Over Social Media Data," *Logistics*, vol. 4, no. 2, p. 12, 2020.
41. Muaad, Abdullah Y., et al., "An Effective Approach for Arabic Document Classification Using Machine Learning," *Global Transitions Proceedings*, vol. 3, no. 1, pp. 267–271, 2022.
42. Rahman, Atiqur and Md Sharif Hossen, "Sentiment Analysis on Movie Review Data Using Machine Learning Approach," *International Conference on Bangla Speech and Language Processing (ICBSLP)*, pp. 1–4, 2019.

6 Efficient Blockchain-Based Edge Computing System Using Transfer Learning Model

P. Sivakumar, S. Nagendra Prabhu,
S.K. Somasundaram, J. Uma Maheswari,
and Murali Murugan

6.1 INTRODUCTION

The IoT devices are integrated in several applications like gesture-tracking devices with a high power and capability. The IoT system has consumed a maximum power because of its broad usage. Many conventional IoT-based cloud communication are performed and redesigned by using a mobile edge computing (MEC) system to reduce computing power. The MEC system is integrated among sensor medium IoT nodes and the cloud environment [1]. Thus, the MEC system achieved a lower computation power using it but the security of the system is required. Decentralized technology like blockchain is integrated with an MEC to achieve anonymity, secrecy, and privacy of IoT cloud data.

Blockchain technology is based on bitcoin and has characteristics of decentralization, transparency, and openness. Generally, this technology has an addition of string data to a blockchain by encryption [2]. This technology is implemented in many confidential sectors like banking, registration, and military data using a trust approach based on credit identification. The blockchain can solve the third-party issue by directly connecting two strange entities. It was a distributed system through consensus mechanisms and node verification [3]. Thus, the distributed blockchain architecture of IoT is presented for a value transfer by transmitting data.

This chapter implements the blockchain-based MEC (BMEC) for IoT data [4]. To achieve a higher security and scheduling strategy, the TL-based clone block identification (CBI) is performed for the detection of clone block attacks. Therefore, the proposed system carried a BMEC with a TL-based CBI to achieve an efficient system. The proposed model's performance is compared with a conventional method. The rest of the work carried a related work in Section 6.2 and the preliminary of TL-based MEC in Section 6.3. Section 6.4 presents the proposed BMEC system, and Section

DOI: 10.1201/9781032623276-6

6.5 includes experimental results. Finally, Section 6.6 includes a conclusion followed by references.

6.2   RELATED WORKS

Samy et al. [4] presented a blockchain system for task offloading security in MEC systems. This method achieved a better performance with better delay and energy consumption. Next, the unmanned aerial vehicle (UAV) system provided secure aerial computing architecture using blockchain, which was mentioned in [5]. This system has ensured the security of computation offloading among mobile users and UAVs. The paper [6] presented by W. Wang et al. that implemented a triple real-time trajectory privacy protection methodology in MEC. W. Wu et al. [7] presented a new approach called the Practical Byzantine Fault Tolerance (PBFT) method for overcoming a committee selection issues and to improve network's security. Next, X. Wang et al. [8] developed a Deep Reinforcement Learning (DRL) based method using a Mean Field Theory (MFT).

In [9], a blockchain-based deep actor-critic method for task offloading is implemented. This method is used to secure the system and solve computation offloading issues. Next, the blockchain-based conditional privacy-preserving authen-tication method is presented in [10]. This system implicated permission for vehicular edge computing. S. Ma et al [11] developed a Byzantine Fault Tolerance (BFT) model to improve the performance. The deterministic network calculus approach is used to analyze the BFT and HotStuff consensus methods.

In [12] a Service Providers location using a blockchain network is presented. This system has various propagation delays when performing computations. P. Kumar et al. [13] developed a mechanism called SAE-ABIGRU to detect Intrusion. This method is used to predict the MEC link load behaviors of the Road Side Unit (RSU) server.

J. Liu et al. [14] proposed a scheme for edge-based networks to increase security and efficiency. They introduce multiple attribute schemes to support a large attribute universe and alleviate management burdens. Edge computing is utilized to speedily respond to vehicle requests and assist resource-constrained computation. Also, to assure the security of users, a cooperative key generation has been used.

X. Huang et al. [15] present a hierarchical federated learning network enabled by a multi-layer blockchain. This allows minimum latency without compromising the security level of the network. Further, the bottleneck analysis has been carried out with imbalanced data distribution.

P. Liu et al. [16] proposed a P2P consortium blockchain-based trading system for resources to incentivize more participants in the context of security and privacy concerns. They focus on the pricing strategy and user concerns, ensuring trust, security on transactions, and privacy of a user without a central authority. C. Xu et al. [1] focus on privacy computing technology and develop a secure architecture based on blockchain.

S. Fugkeaw et al. [17] suggest an IoT aggregation and data transmission method using digital signing and encryption, and they also highlight the collaboration between fog nodes and blockchain. They introduce an encryption method combined with a privacy-preserving access policy and a lightweight policy update algorithm for the effective management of data. Comparative analysis and experiments demonstrate the scheme's computation cost and performance. The hybrid system is proposed by Y. Zhong et al. [18] to design a double-layer mining service offloading. The proposed system combines MEC with a blockchain. The task of offloading is handled by the algorithm of actor-critic (A3C) algorithm and double auction.

Ren et al. [19] have introduced a task-offloading strategy known as blockchain-based trust-aware task offloading strategy, which leverages blockchain technology and trust-awareness. The authors employ smart contracts to facilitate secure and automated task offloading. To mitigate network insider attacks, they incorporate a recommendation filtering technique and a trust penalty measure. Furthermore, the researchers analyze the challenges related to time efficiency when integrating blockchain into their proposed system.

In [20], a blockchain authentication scheme is proposed for securing the messages. The authors used edge servers and a lightweight message authentication algorithm, reducing the computational overhead. Du et al. [21] present a system called Blockchain-based Intelligent Edge Cooperation System (BIECS), which combines blockchain and intelligent edge cooperation to enable efficient resource sharing.

Tian et al. [22] introduce Blockchain-based Secure Searching for Metadata (BSSMeta), a secure searching approach for metadata in MEC using blockchain technology. The authors employ various techniques such as lightweight proxy re-encryption, master-slave smart contracts, and buffer uploading. These measures enable efficient searching and uploading of metadata within the system. Chavhan et al. [23] present a security scheme for vehicle platooning that integrates edge computing and blockchain technology. The authors designed a security architecture and implemented smart contracts that were seamlessly integrated with the Simulation of Urban Mobility Traffic Control Interface Application Programming Interface. This integration ensures the establishment of a secure environment for vehicle platooning.

L. Qi et al. [24] present a cooperative proof-of-work (PoW) strategy named Relay-PoW, which reduces the consumption of energy and improves the efficiency of resource utilization. They proposed a supervision group mechanism and parallel relay mining to increase the throughput and ensure security. B. Jiang et al. [25] propose an attribute-based encryption technique, blockchain, and smart contracts for access control and event traceability.

Lang et al. [26] introduce a framework for cooperative carbon monoxide (CO) detection and utilize blockchain technology for secure handover in Vehicle Edge Computing. The authors incorporated a consensus mechanism to develop models for vehicle mobility and cooperative handover to guarantee the immutability of offloaded data. J. Li et al. [27] propose a blockchain-assisted access control technique for healthcare systems in fog computing. They utilize smart contracts and a multi-authority attribute-based encryption scheme to ensure data transparency, reliability, and confidentiality.

Y. Li et al. [28] propose a blockchain-based data integrity verification scheme for cloud-edge computing environments. They propose a detailed scheme without a trusted third party, consisting of data upload and verification phases. R. Li et al. [29] propose a secure blockchain-based access control framework for the Industrial

Internet of Things (IIoT). They design an access frequency-based network sharding scheme to improve scalability and a Bloom filters-based privacy protection scheme.

X. Ding et al. [30] consider the case where IoT devices require assistance from edge servers for data analysis and storage-related tasks, which affect their profits based on the resources purchased. They propose an allocation strategy where IoT devices can purchase resources from various edge servers, and they can allocate a limited budget to purchase the resources to increase the profit. In their study, L. Sun et al. [31] address the heterogeneous and dynamic nature of fog nodes (FNs) in fog computing resource management. They propose a truthful auction for the fog system (TAFS) to incentivize FNs to satisfy application demands with guaranteed performance. It considers both delay tolerance and resource availability for all applications.

Du et al. [32] introduce a matching mechanism based on smart contracts aimed at establishing renting associations between Distribution System Operators (DSOs) and Energy Consumption Nodes (ECNs) with the goal of maximizing social welfare. Additionally, the authors introduce a double auction technique to calculate the pricing for the winning applicants.

In their work, Nguyen et al. [33] propose a proof-of-reputation consensus mechanism based on a block verification strategy. Their research focuses on developing a multi-objective function within the BMEC system. This function aims to maximize the system utility by optimizing various aspects, including offloading decisions, resource, and power allotments.

6.3 PRELIMINARIES

In this section, the TL-based MEC is presented for an IoT application. This TL model is based on the Convolutional Neural Network (CNN) structure of the DL approach. The TL model is applied for fast computing, and it allocates a minimum amount of time to learn the huge dataset. This TL model achieved a robustness of the system and also computational delay is reduced in it due to its shortcut.

TL Model structure
In this MEC system, the VGG16-TL model is used for IoT data computing. The VGG16 model provided a better system performance with a higher accuracy but does not reduce the learning device.

VGG16 Description
The VGG16 is applied to a large-scale dataset that provides an ImageNet. This model reduced the delay of learning time and also minimized insufficient data. Figure 6.1 shows the VGG16 block diagram that consists of a total of 16 layers, including 13 convolutional layers and 3 dense layers.

From Figure 6.1, the VGG16 consisted of convolutional and dense layers. The dataset is used to learn the dense layer weights. Several convolutional layers are categorized, namely 64 filters in 2 layers, 128 filters in 2 layers, 256 filters in 3 layers, and 512 filters in 3 layers. All the convolutional filters provided a 3×3 size with a

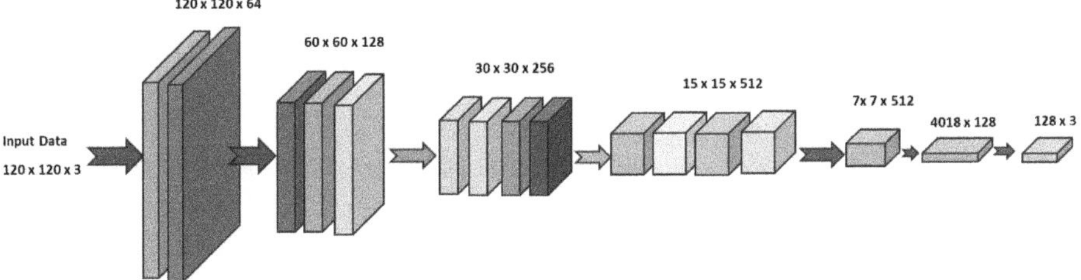

FIGURE 6.1 VGG16 Transfer-learning models.

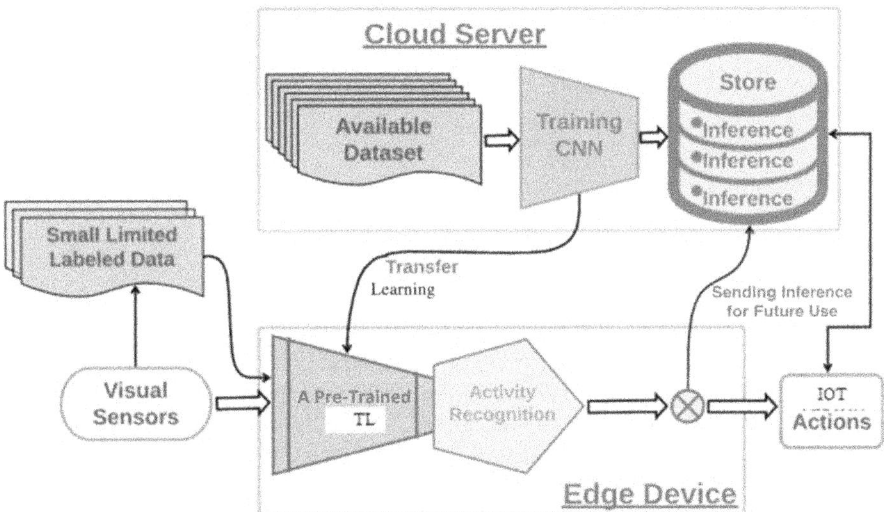

FIGURE 6.2 Architecture of TL-based MEC in IOT.

maximum pooling layer of 2×2 size. This VGG16 has two dense layers such as one as a hidden layer with 128 neurons and another as an output layer. There are approximately 138 million parameters provided for the network.

TL-Based MEC in IOT

The VGG16 TL model is applied to IoT data in the MEC system, as shown in Figure 6.2. The datasets are learned by using a Graphics Processing Unit (GPU)-enabled cloud in this system. The trained dataset is processed by the Edge server and provided with image data using visual sensors. The fine-tunings are performed with a small amount of ground-labeled data in this system. Next, the edge server processes edge computing using an installed sensor. The activity of an edge server is recognized using a VGG16 classifier. The predicted inference is sent to the cloud server. Then, a cloud server takes appropriate action based on present inference and stored information.

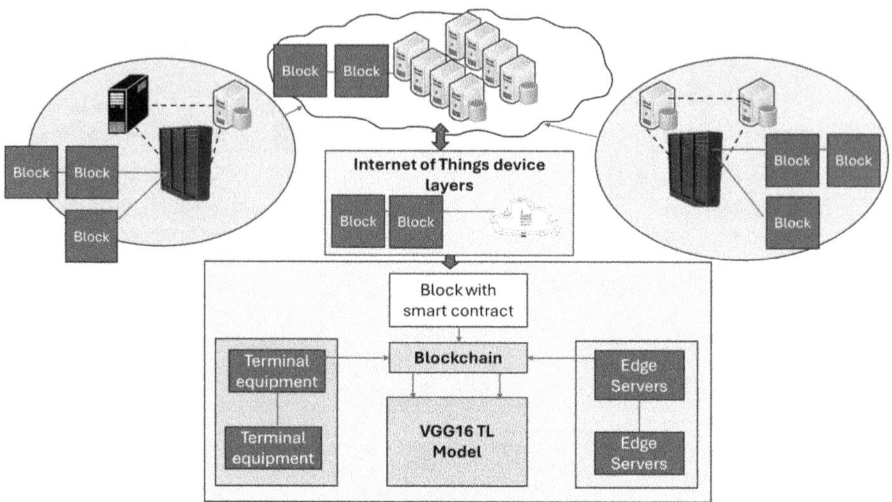

FIGURE 6.3 Proposed architecture of BMEC system.

6.4 PROPOSED SYSTEM

The VGG16 TL method was efficient for a MEC-IoT system due to its minimum delay in scheduling time. However, the security of an IoT data in the MEC system is its biggest threat. To obtain security, the blockchain technology is integrated to TL-based MEC system. Therefore, the proposed architecture of the MEC system based on the blockchain is shown in Figure 6.3. This system comprised a mobile terminal device, cloud servers, edge servers, and blockchain.

The IoT provides a mobile terminal device that is integrated with a block with smart contracts. In this layer, the BMEC generated a blockchain account. The edge servers are interconnected in the edge network layer. These layers have specific blockchains on the edge layer and are also used to monitor the BMEC transaction. Whenever blockchain transactions are processed, new blocks are created. The cloud servers are interconnected and provide a blockchain network. The proposed system can achieve a speedy transaction in the cloud computing layer by optimizing the blockchain topology.

TL-Based Clone Block Identification

To enhance security, clone block attack detection is the major concern in the BMEC system. The clone attack of the block was an illegal attacker. This attack imitates the authorized block and gathers its legal information and copied to another blocks. The original ID and hash key verification data are cloned in various locations in the network. Figure 6.4 shows the clone block attack architecture that consists of an edge node, a legal block, and three clone blocks. The clone blocks I, II, and III have fake information about the original block. The fake content in clone block I is greater than clone block 2 and clone block 3. Clone block II fake content is larger than clone block III.

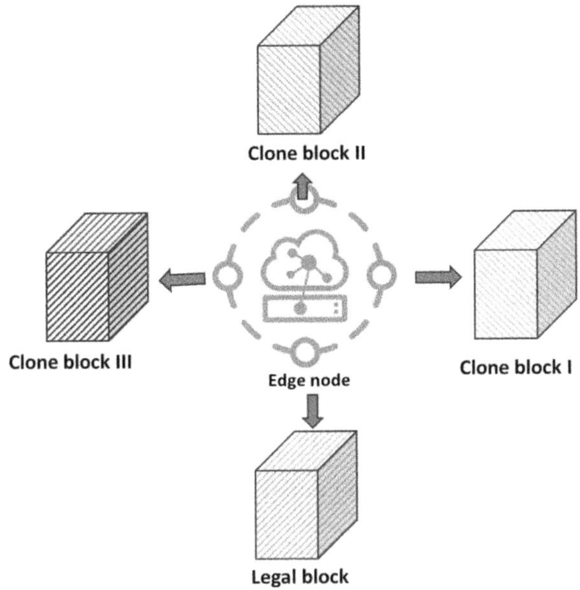

FIGURE 6.4   Clone blocks attack.

To identify this issue, the CBI is performed using a VGG16 TL model shown in Figure 6.4. The TL model is very effective in prediction by training a few data sets. The clone blocks are identified with their higher classification and prediction strategy. This proved that the proposed TL-based CBI can be used to improve security in the BMEC system. Therefore, the proposed BMEC system achieved both scheduling accuracy and security using TL-based CBI.

VGG Tuning

The parameters in VGG16 include the weights and biases of these layers, which are learned through the training process. These parameters define the network's ability to extract features from input data and make accurate predictions.

In deep learning, tuning of parameters is an essential step for the performance optimization of a model. It involves adjusting the values of the model's parameters to achieve better accuracy or other desired metrics. Grey Wolf Optimization is a metaheuristic algorithm based on the social behavior of wolves. It mimics the hunting behavior of gray wolves to find the optimal solutions.

The objective function could be a combination of different metrics that capture the system's performance, such as classification accuracy, system delay, and security measures. For example, you could define the objective function as a weighted sum of accuracy and delay, where accuracy represents the model's performance in correctly classifying data, and delay represents the time taken for processing.

The Grey Wolf Optimization (GWO) algorithm would then be utilized to search the parameter space of VGG16 and find the optimal values that maximize the objective function. GWO is a population-based algorithm that emulates the hunting behavior of gray wolves, consisting of alpha, delta, beta, and omega wolves representing different positions in the search space.

During the optimization process, each wolf in the GWO algorithm corresponds to a set of parameters in VGG16. The position of a wolf represents a potential solution in the parameter space. By iteratively updating the positions of wolves based on their hunting behavior, GWO explores the parameter space and gradually converges toward the optimal solution.

The behavior of wolves in GWO involves three main steps: encircling, attacking, and updating positions. These steps can be related to the optimization process as follows:

Encircling: In this step, the GWO algorithm performs a search around the current positions to explore new potential solutions. This simulates the exploration of different parameter combinations in VGG16.

Attacking: The GWO algorithm identifies the most promising positions and updates them to move closer to the optimal solution. This step corresponds to refining the parameter values of VGG16.

Updating positions: The GWO algorithm changes the positions of wolves based on their hunting behavior and the positions of the alpha, beta, delta, and omega wolves. This step helps the algorithm converge toward the optimal solution and fine-tune the parameters of VGG16.

Applying GWO-based tuning to VGG16, the objective function defined earlier guides the optimization process to find the parameter values that maximize the desired goals of efficient security and lower delay in the BMEC system. The GWO algorithm explores the parameter space of VGG16, updating the weights and biases, until an optimal solution is found. The pseudocode for the proposed GWO-based parameter tuning is given below:

- Initialize a population of wolves (alpha, beta, delta, omega)
- Initialize VGG16 with random parameter values
- Define the objective function
- Function objective_function (parameters)
- Load VGG16 with the given parameters
- Train VGG16 on a labeled dataset
- Evaluate VGG16's performance on a validation set
- Compute the objective function value based on the desired metrics
- Return objective_value
- Initialize the maximum number of iterations
- Initialize convergence criteria
- While not converged and iterations < max_iterations
- For each wolf in the population
- Encode the parameters of the wolf into VGG16
- Compute the objective function value using objective_function()
- Sort the population based on the objective function values

- Update positions of the wolves
- For each wolf in the population
- Update the positions based on the hunting behavior of wolves
- Perform parameter adjustments
- For each wolf in the population
- Adjust the parameters in VGG16 based on the positions of the wolves
- Increment the iteration counter
- Select the best wolf with the highest objective function value
- Retrieve the parameters from the best wolf
- Return the tuned parameters of VGG16

Through this iterative optimization process, the tuned VGG16 model achieves improved performance, leading to better scheduling strategies, efficient security, and reduced delay in the BMEC-IoT system compared to conventional methods. The transfer learning (TL) based CBI further leverages the optimized VGG16 model to improve the overall efficiency and effectiveness. The GWO algorithm explores the parameter space and refines the parameters based on the hunting behavior of wolves. The tuned VGG16 model contributes to the improved performance of the BMEC-IoT system, resulting in better scheduling.

6.5 EXPERIMENTAL RESULTS

This section presented an experimental result of the proposed BMEC system. At first, consider ten blocks for blockchain sharing. Assign the connection weight among every block randomly. In this method, the TL-based CBI is used to recognize the next block generation. From Figure 6.5, the blocks in the edge server are directly

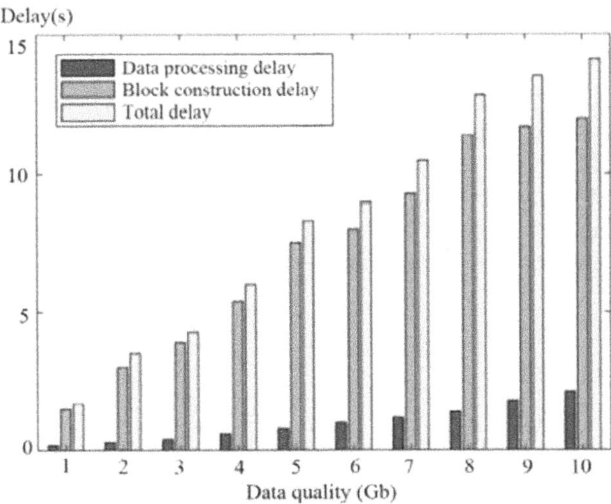

FIGURE 6.5   Total delay in proposed BMEC system.

TABLE 6.1
Performance Table of Scheduling Accuracy

Methods	Scheduling Accuracy
TL-based MEC	98.76
Proposed BMEC	99.54

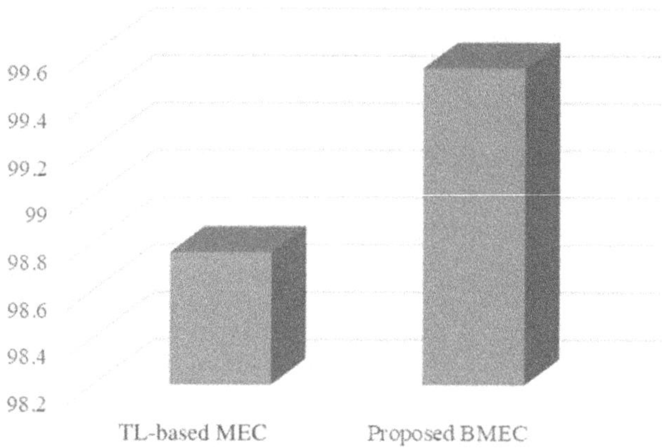

FIGURE 6.6   Performance of scheduling accuracy.

proportional to the total delay in BMEC, where both the parameters increase non-linearly. This showed that the system can enhance the throughput transaction and have low latency of computation.

Table 6.1 and Figure 6.6 show the proposed BMEC and TL-based MEC performances in scheduling accuracy. The proposed method has achieved a 99.54% and TL-based MEC attained 98.76% scheduling accuracy. The result showed that the proposed BMEC method is much better than the existing system. Therefore, the proposed system has obtained both scheduling accuracy and security by reducing computational delay. This shows that system is more scalable than conventional systems.

6.6   CONCLUSION

In this work, the BMEC system is proposed and designed using a TL-based CBI method. The TL-based CBI method is used to identify the clone block attack that was attained in the BMEC system. This system consists of IoT devices, an Edge server, a Cloud Server, and a Blockchain. The implementation of blockchain achieved a higher security with a TL-based CBI method. This method predicted the various clone block attacks in the edge server with a knowledge of legal block ID and Key functions. The performance result showed that the total delay in BMEC gets increased non-linearly, which achieved a high throughput transaction and low latency. Also, the proposed BMEC system has obtained a scheduling accuracy of 99.54% compared to the existing TL-based MEC system. This showed that this system is more scalable than the conventional systems.

REFERENCES

[1] C. Xu, X. Yu, H. Cui, H. Qiu, Y. Huang, and Y. Xiao, "Privacy computing technology and secure architecture based on blockchain," *Journal of Parallel and Distributed Computing*, vol. 145, pp. 68–80, 2020.

[2] J. Liu, T. Peng, and W. Wang, "Blockchain-based edge computing resource allocation in IoT: A deep reinforcement learning approach," *IEEE Internet of Things Journal*, vol. 8, no. 6, pp. 4099–4108, 2021.

[3] W. Jin, Y. Xu, Y. Dai, and Y. Xu, "Blockchain-based continuous knowledge transfer in decentralized edge computing architecture," *Electronics*, vol. 12, no. 5, p. 1154, 2023.

[4] K. Samy, Y. R. Rezk, M. A. R. Ghany, and A. A. R. Shehab, "A blockchain system for task offloading security in MEC systems," *Journal of Network and Computer Applications*, vol. 158, p. 102660, 2020.

[5] Q. Tang, H. Zhu, Y. Luo, and J. Liu, "Secure aerial computing architecture using blockchain," *IEEE Transactions on Network and Service Management*, vol. 16, no. 3, pp. 1047–1058, 2019.

[6] W. Wang, X. Li, X. Shen, and Y. Liu, "Triple real-time trajectory privacy protection methodology in MEC," *IEEE Transactions on Mobile Computing*, vol. 20, no. 1, pp. 101–115, 2021.

[7] W. Wu, X. Yuan, Y. Zhang, and J. Zhang, "PBFT method for overcoming committee selection issues and improving network security," *IEEE Transactions on Parallel and Distributed Systems*, vol. 32, no. 5, pp. 1034–1047, 2021.

[8] X. Wang, Y. Wang, Z. Wang, and Y. Sun, "Deep reinforcement learning (DRL) based method using mean field theory (MFT)," *IEEE Transactions on Neural Networks and Learning Systems*, vol. 31, no. 8, pp. 2901–2913, 2020.

[9] S. Zhang, X. Wang, T. Guo, and Z. Huang, "Blockchain-based deep actor-critic method for task offloading," *IEEE Access*, vol. 9, pp. 158205–158218, 2021.

[10] J. Yang, X. Wei, Z. Wang, and H. Zhao, "Blockchain-based conditional privacy-preserving authentication method for vehicular edge computing," *IEEE Internet of Things Journal*, vol. 8, no. 3, pp. 1831–1841, 2021.

[11] S. Ma, T. Zhang, Y. Xu, and W. Wu, "BFT model to improve performance in MEC," *IEEE Transactions on Computers*, vol. 70, no. 7, pp. 1085–1099, 2021.

[12] S. Huang, X. Liu, J. Chen, and Q. Wu, "Service provider location using a blockchain network," *Sensors*, vol. 21, no. 5, p. 1652, 2021.

[13] P. Kumar, S. Singh, R. Misra, and Y. Liu, "SAE-ABIGRU: A mechanism for intrusion detection in MEC," *IEEE Transactions on Information Forensics and Security*, vol. 16, pp. 3005–3016, 2021.

[14] J. Liu, Z. Zhu, X. Hu, and Y. Zhang, "Edge-based networks for increased security and efficiency," *IEEE Transactions on Dependable and Secure Computing*, vol. 18, no. 4, pp. 1770–1781, 2021.

[15] X. Huang, Y. Liu, W. Wang, and T. Zhang, "Hierarchical federated learning network enabled by a multi-layer blockchain," *IEEE Transactions on Network and Service Management*, vol. 18, no. 1, pp. 100–113, 2021.

[16] P. Liu, L. Zhang, J. Zhou, and S. Wang, "P2P consortium blockchain-based trading system for resources," *IEEE Transactions on Emerging Topics in Computing*, vol. 9, no. 2, pp. 769–782, 2021.

[17] S. Fugkeaw, T. Boonkrong, P. Jantaporn, and A. Sae-Tang, "IoT aggregation and data transmission method using digital signing and encryption," *Future Generation Computer Systems*, vol. 116, pp. 164–175, 2021.

[18] Y. Zhong, Q. Wang, M. Li, and X. Tang, "A hybrid system for double-layer mining service offloading," *IEEE Transactions on Cloud Computing*, vol. 8, no. 1, pp. 14–27, 2020.

[19] J. Ren, Y. Zhang, X. Huang, and T. Wang, "Task-offloading strategy leveraging blockchain technology and trust-awareness," *IEEE Transactions on Sustainable Computing*, vol. 6, no. 4, pp. 593–606, 2021.

[20] F. Wang, Z. Liang, T. Chen, and Y. Sun, "Blockchain authentication scheme for messages," *Journal of Network and Computer Applications*, vol. 173, p. 102881, 2021.

[21] X. Du, L. Chen, Y. Liu, and R. Wang, "BIECS: Blockchain and intelligent edge cooperation for resource sharing," *IEEE Internet of Things Journal*, vol. 8, no. 10, pp. 8555–8567, 2021.

[22] Y. Tian, Z. He, L. Li, and Y. Xu, "BSSMeta: A secure searching approach for metadata in MEC using blockchain," *IEEE Transactions on Services Computing*, vol. 14, no. 4, pp. 1371–1384, 2021.

[23] A. Chavhan, S. Patil, Y. Fang, and R. Zheng, "Security scheme for vehicle platooning integrating edge computing and blockchain," *IEEE Transactions on Intelligent Transportation Systems*, vol. 23, no. 3, pp. 2034–2046, 2021.

[24] L. Qi, X. Liu, Z. Wang, and Y. Wu, "Cooperative proof-of-work (PoW) strategy named Relay-PoW," *IEEE Transactions on Information Theory*, vol. 67, no. 4, pp. 2323–2338, 2021.

[25] B. Jiang, Q. Zhao, Y. Liang, and M. Li, "Attribute-based encryption technique, blockchain, and smart contracts for access control," *IEEE Access*, vol. 9, pp. 184123–184137, 2021.

[26] C. Lang, Y. Huang, Z. Zhang, and S. Wei, "Cooperative carbon monoxide detection using blockchain technology," *IEEE Transactions on Vehicular Technology*, vol. 70, no. 8, pp. 7860–7870, 2021.

[27] J. Li, X. He, Y. Zhang, and M. Chen, "Blockchain-assisted access control for healthcare systems in fog computing," *IEEE Transactions on Emerging Topics in Computing*, vol. 9, no. 3, pp. 1264–1275, 2021.

[28] Y. Li, H. Zhao, X. Xu, and Y. Wang, "Blockchain-based data integrity verification scheme for cloud-edge computing," *Future Generation Computer Systems*, vol. 117, pp. 341–351, 2021.

[29] R. Li, W. Yu, H. Zhao, and Y. Sun, "Secure blockchain-based access control framework for IIoT," *IEEE Internet of Things Journal*, vol. 8, no. 4, pp. 2623–2636, 2021.

[30] X. Ding, Y. Liu, L. Li, and J. Yang, "IoT devices assistance from edge servers for data analysis and storage-related tasks," *IEEE Transactions on Cloud Computing*, vol. 9, no. 4, pp. 1323–1334, 2021.

[31] L. Sun, X. Zhou, T. Chen, and Y. Gao, "Truthful auction for fog system (TAFS) to incentivize fog nodes," *IEEE Transactions on Network and Service Management*, vol. 18, no. 2, pp. 1884–1896, 2021.

[32] X. Du, Y. Wang, Q. Zhang, and L. Chen, "Matching mechanism based on smart contracts for renting associations," *IEEE Transactions on Smart Grid*, vol. 12, no. 1, pp. 567–578, 2021.

[33] H. Nguyen, X. Wang, T. Zhang, and P. Li, "Proof-of-reputation consensus mechanism based on block verification," *IEEE Transactions on Network and Service Management*, vol. 18, no. 1, pp. 104–116, 2021.

Introducing a Compact and High-Speed Machine Learning Accelerator for IoT-Enabled Health Monitoring Systems

Kavitha V.P., Magesh V., Theivanathan G., and S.S. Saravana Kumar

7.1 INTRODUCTION

Recently, high-speed ML accelerators have evolved into new innovations due to the arrival of IoT-enabled AI integration [1]. Recent trends in medical applications have seen a paradigm shift toward edge and embedded devices, emphasizing the need for high-performance hardware to meet the growing demands of real-time data processing. As the Internet of Things (IoT) continues to expand its footprint across industries, the integration of artificial intelligence (AI) at the edge has become a critical driver of progress [2].

High-speed machine learning accelerators are used in different fields like image and signal processing. The hardware components in accelerators support for the integration of AI and ML programs [3]. Also, it allows an IoT environment for continuous monitoring. The data is updated to the cloud continuously with the consumption of huge amount of power. The collected data from the cloud is used for different forecasting, such as weather, disaster, traffic, etc.

Recent trends in the tech industry highlight the increasing prevalence of IoT ecosystems across domains such as health care, smart cities, industrial automation, and autonomous vehicles. These trends make the need for efficient and high-speed ML acceleration at the edge. The processors needed to work heavily when the volume of the data increases [4]. As data volumes grow exponentially, the ability to process and interpret this data directly on the device offers substantial advantages, including reduced latency, enhanced privacy, and more robust operation, particularly in environments with limited connectivity.

In this work, design optimization of a High-Speed Machine Learning Accelerator is proposed, which brings AI capabilities closer to the edge in IoT deployments. By using cutting-edge techniques such as Support Vector Machines (SVM) for data

DOI: 10.1201/9781032623276-7

truncation and Honey Batcher optimization for hyperparameter tuning, our hardware accelerator promises to be a game-changer in the IoT landscape. The performance of the processors can be improved in terms of accuracy and efficiency for various applications.

This work also explains optimization methodologies to illustrate how these advancements contribute to the realization of intelligent and responsive IoT devices. By applying the high-speed ML revolution, we prove that our work will give significant advancements in IoT-enabled AI integration, opening up new possibilities for innovation and problem-solving across multiple sectors.

7.2 RELATED WORK

Ali et al. [5] analyzed the performance of accelerators with different ML algorithms. By varying the input size, the power consumption of the processors was analyzed using various algorithms. The processor with a Field Programmable Gate Array (FPGA) board shows superior performance in terms of delay and area.

The concept of prediction is used in many fields for future planning. The prediction-based processor is designed by Braatz et al. [6]. The delay of a particular algorithm is detected before the execution to save the power. The resources are allocated earlier to reduce the overall delay of the circuits. Lee and Han [7] constructed a matrix optimization-based hardware accelerator for low-power applications. The busy and idle states of the program are calculated, and the functions of the matrix are optimized with array segmentations.

Gentile and Serio [8] explored the concept of a new feature extraction algorithm for the design of a big data processing hardware accelerator. The data from a huge database is processed using a new feature extraction algorithm, and relevant data is identified to avoid unnecessary processing of data. In Kee et al. [9], the author designed a SVM-based hardware accelerator for environment processing. The sensor data collected from the environment is processed by ML algorithms, and redundant data is removed to avoid redundant storing of data. Experimental results show that the power savings achieved up to 5% when applying ML.

Chen and Hao [10] proposed a neural network-based data fusion algorithm for hardware accelerators. The neurons are processed to locate redundant data in the processor, and a masking strategy is applied to eliminate unnecessary data. Song et al. [11] developed an image-processing hardware accelerator for high-speed applications. Then, the FPGA processor is used for image processing and analyzed in terms of slice, look tables, and delay. Finally, the image segmentation algorithm is loaded for power evaluations.

The concept of approximate computing is used to achieve low power in processors. The approximate messaging approach was applied in communication systems by Brennsteiner et al. [12] to achieve a low power. The data is handled effectively with minimum error rates than other algorithms. Opportunistic redundancy is the approach used in communication systems to improve data transmission reliability over unreliable or error-prone channels. The opportunistic redundancy-based hardware accelerator is designed by Dong et al. [13] for signal processing. The overall area and time requirements are optimized by the proposed approach.

Lai et al. [14] proposed a new attack detection mechanism for hardware accelerators. The ML-trained mechanism is utilized to learn the attacker's behavior when compared to the normal users. The data from the Kaggle website is used to train the ML model for attacker detection.

Zheng et al. [15] applied encryption algorithms in ML-based hardware accelerators. The security of the processors can be improved by applying encryption algorithms in accelerators. The homomorphic encryption is processed to secure data in accelerators to prevent attackers'-learning is a reinforcement learning algorithm used in ML. It's designed to solve problems with minimum computational requirements. The concept of q learning used in hardware accelerators by Sutisna et al. [16] for multimedia processing applications.

Wang et al. [17] constructed a hardware accelerator using the strategy of parallelism. The parallelism is used to minimize the execution time by applying simultaneous execution. The data is processed parallelly to avoid latency in execution. Aboye et al. [18] proposed a new processor design using triple modular redundancy. The fault-tolerant capability of the system can be improved by placing redundant modules. These redundant modules help when an error is detected.

The reconfiguration of the system is used to switch over the functionality based on user requirements. The reconfiguration-based accelerator is proposed by Vranjković and Struharik [19] for multimedia applications. The power consumption of the processor can be optimized by reconfiguration. Edge-enabled ML refers to the practice of deploying machine learning models and algorithms on edge devices. This approach was applied by Suresh and Renu Madhavi [20] in hardware accelerators for communications systems.

The hybrid Polyvalent ML-based hardware accelerator is proposed by Zhou et al. [21]. It combines the different ML algorithms for processing the data in accelerators. Based on applications, the ML algorithm can be loaded from memory. Aizaz et al. [22] designed a new type of multiplier architecture for ML-based hardware accelerates. The power consumption of processors is greatly optimized by introducing new multiplier structures. Likewise, Nurvitadhi et al. [23] proposed a new vector multiplication architecture for ML-based accelerator design. Xin et al. [24] developed a new multi-layer architecture for image-processing ML hardware accelerators. The pixels in the processors are optimized using the multi-layer network.

7.3 PRELIMINARY

7.3.1 MULTIPLIERS

High-speed machine learning accelerators have become essential part all fields for real-time AI integration and processing at the edge. The multiplier is the key component of accelerators, which performs data multiplications and addresses calculations.

Multipliers are specialized hardware units designed to perform the crucial task of rapidly and efficiently executing complex mathematical operations. It plays a major role in ML algorithms for difficult computations. The demand for high-speed and low-power multipliers is increases when ML-based applications grow in all fields.

For high-speed machine learning accelerators, multipliers are the key components responsible for executing the numerous matrix multiplications and convolution operations essential for tasks such as image processing, NLP, pattern recognition, and analysis. The performance of multipliers directly impacts the overall speed and efficiency of an ML-based accelerator. Different algorithms and architecture have been introduced by the researchers to design a multiplier in minimal area and delay. The integration of the inelegance algorithm in the multiplier leads to area and delay minimization achievements.

7.3.2   HBO

The honey badger's remarkable foraging behavior serves as the inspiration for the Honey Badger Optimization (HBO) model, a nature-inspired algorithm that aims to solve complex optimization problems. This mathematical model is based on the two key phases observed in the honey badger's foraging process: the digging and honey stages.

Initialization (Step 1): The HBO model begins by initializing a population of candidate solutions. The population is represented as an array, where each element corresponds to a potential solution. Each solution is represented as a vector, and the population initialization involves generating these vectors within specified bounds.

Candidate Solution Population (pop):

$$population\left(pop\right) = \left[X_{11}\ X_{12}\ldots X_{21:}\ X_{22:}\ldots X_{n1}\ X_{n2}\ldots\quad X_{1P}\ X_{2P:}\ X_{nP}\right] \tag{7.1}$$

$$\text{Honey badger } 1^{\text{th}}\text{position}, x_1 = \left[x_l^1, x_l^2, \ldots, x_l^P\right] \tag{7.2}$$

In above equation, "pop" represents the entire population, and each "x_l" in Equation (7.3) is an individual candidate solution as follows:

$$x_l = lb_l + RN_1 \times \left(ub_l - lb_l\right) \tag{7.3}$$

where *RN* is the random number with upper and lower bounds.

Intensity Calculation (Step 2): In the HBO model, an intensity value (IV) is calculated based on Inverse Square Law. This value represents the energy or distance between the badger to prey.

Square Law expressed in Equation (7.4).

$$IV_l = RN_2 \times \frac{S}{4\pi d_l^2} \tag{7.4}$$

where

$S = \left(x_l - x_{l+1}\right)^2$, that indicated the concentration strength

$d_l = x_{prey} - x_l$, that specified a distance among prey and l^{th} honey badger

Here, "S" represents the concentration strength, and "d_l" is the distance from source to destination.

Density Factor Update (Step 3): α-density factor controls the time-varying randomization, facilitating balance between convergence rate during the optimization approach. This factor decreases over time, smoothing the solution.

$$\propto = C \times expexp\left(\frac{-t}{t_{max}}\right) \tag{7.5}$$

where

$t_{max} \rightarrow$ allowable computation

$C \rightarrow$ Constant ≥ 1 (default $= 2$)

In above Equation, "t_max" is the Max_No. of iterations, and "C" is a constant typically set to a value greater than or equal to 1.

Escaping Local Optima (Step 4): The HBO model introduces a mechanism to escape local optima. A flag (F) is used to change the position, enabling a more extensive exploration of the search space.

Position Update (Step 5): The location of badger is updated using the digging stage and the honey stages.

$$X_{new} = X_{prey} + F \times \beta \times I \times X_{prey} + F \times r_3 \times \propto \times d_l \times \left| cos\left(2\pi.rand_4\right) \times \left[1 - coscos\left(2\pi.rand_5\right)\right]\right| \tag{7.6}$$

where

$X_{prey} \rightarrow$ prey position

$\beta \rightarrow$ ability factor of Honey Badger

There is a flag (F) work is used to modify the direction of search that is expressed in below Equation (7.12):

$$F = \{1 \quad if \ rand_6 \leq 0.5 - 1 \quad else, \tag{7.7}$$

In the phase of Honey stage, the beehive is attained by a honeyguide bird using honey badger which is expressed as follows (7.8):

$$X_{new} = X_{prey} + F.rand_7. \propto .d_l \tag{7.8}$$

In above Equation, *"X_new"* represents the new position, the position of the prey denoted as *"X_prey"*, *"F"* is flag controlling the search direction, *"β"* represents ability of honey_badger's, *"I"* the intensity, *"d_I"* is the distance between best solutions, and various random values are used for perturbation.

The HBO model encompasses these phases, combining exploration and exploitation to find optimal solutions in the search space. By adapting the behavior of the honey badger in foraging for solutions, the HBO model provides a unique and efficient approach to solving complex optimization problems.

7.3.3   SVM

SVM are a powerful ML approach especially used for classification problems. SVMs are mainly used where data points belong to two distinct classes and need to be separated by a hyper_plane. The basic strategy of SVM, is to discover the most advantageous hyper_plane that optimizes the distinction between the two classes, leading to improved generalization and robustness.

The objective of SVM is to find the decision boundary that separates the data into two classes. This decision boundary is represented by a hyperplane, which can be expressed as:

$$\text{Hyper\_plane Equation: } w * x + b = 0 \tag{7.9}$$

where, *"w"*-weights or coefficients of the hyper_plane, *"x"*-feature vector of the data point, *"b"*-bias term, and *"f(x)"* decision function.
The decision function can be defined as follows:

$$\text{Decision Function: } f(x) = \text{sign}(w * x + b) \tag{7.10}$$

In above Equation, *"f(x)"* is the decision function, "sign" is the sign function (indicates the fall of data point on the hyper_plane), and *"b"* is the bias term.

7.4   PROPOSED SYSTEM

This work proposes a new ML-based multiplier to improve the computational efficiency of health care data processing while maintaining accuracy. The proposed architecture also supports medical image processing. The proposed architecture is best suited for applications where power consumption and resource utilization are critical concerns. It combines two key elements: multiplier truncation and compensation values and the selection of optimal values using SVM, as shown in Figure 7.1.

Multiplier Truncation and Compensation
This aspect of the system involves the controlled reduction of the precision of multipliers. The truncation of less significant bits of the multiplier leads to lower computations and reduced power computations. However, this truncation introduces errors in the results. The compensation circuits add compensation values to reduce

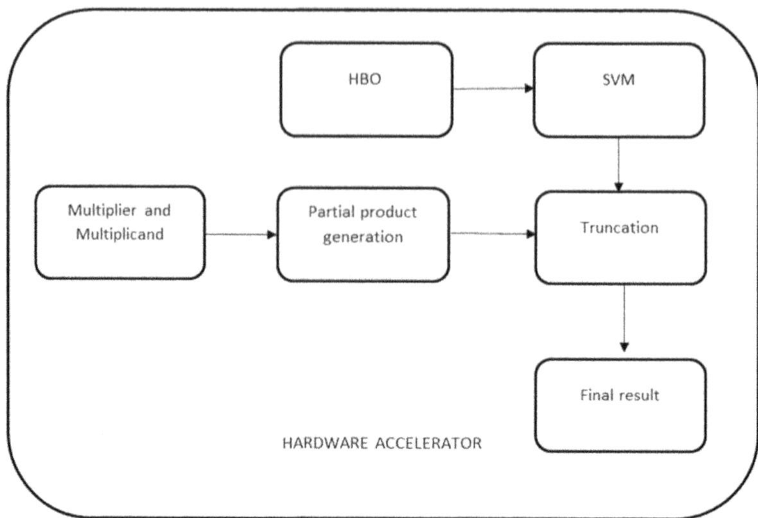

FIGURE 7.1 Proposed architecture for disease prediction.

that error. Compensation values are introduced to address the accuracy loss caused by multiplier truncation. These values are determined through optimization techniques with the goal of minimizing the impact of truncation errors. The choice of these compensation values is critical to balancing efficiency and precision.

Here, the SVM is used to find out the best compensation values in order to reduce an error value of multiplier during truncation. The error values of the multiplier changed based on the number of ones present in the least significant bits of the multiplier output.

The system utilizes approximate multipliers in its architecture. Approximate multipliers are designed to provide fast but slightly less precise results. These multipliers are well-suited for applications where a small degree of error is acceptable in exchange for significant improvements in computational speed and efficiency.

The design of a truncated multiplier includes three different stages: partial product generation, truncation of bits, and final vector merging addition stages. In truncation, the hyperparameter-tuned SVM is applied for truncation and adds compensation values to the multiplier final outputs.

By using approximate multipliers in hardware accelerators, the system achieves a substantial improvement in computational efficiency. The proposed architecture is best suited for ML-based hardware accelerators, especially in real-time signal processing, edge computing, and IoT devices.

The reduced precision of multipliers and optimized compensation values lead to savings in terms of hardware resources and power consumption. This is particularly valuable in resource-constrained environments. While focusing on efficiency, the system ensures that a satisfactory level of accuracy is maintained through the careful selection of compensation values. This is achieved with the assistance of the SVM, which helps strike a balance between computational speed and precision.

Hyperparameters of SVM

SVMs offer several hyperparameters that allow for fine-tuning the model to achieve the best performance for a specific dataset. The primary hyperparameters include:

Kernel Function (K): SVMs can use different kernel functions to formulate data into a large-dimensional space. The SVM kernel functions are classified as linear, radial basis function (RBF), polynomial and sigmoid.

Regularization Parameter (C): The regularization parameter "C" reduces the classification error. The minimum value of "C" denotes a wider margin but may lead to errors, while a larger "C" emphasizes accurate classification, potentially at the cost of a narrower margin.

Kernel-specific Hyperparameters: If a kernel function is used (e.g., the RBF kernel), there are additional hyperparameters specific to the kernel. For the RBF kernel, these include "gamma," which controls the shape of the decision boundary.

HBO-Based Hyperparameter Tuning

The HBO algorithm is a nature-inspired optimization technique that can be applied to SVM hyperparameter tuning. HBO leverages the behavior of honey badgers in foraging for prey to explore and exploit the search space for optimal solutions.

In the context of SVM hyperparameter tuning, HBO can be used to find the best combination of hyperparameters, such as the choice of kernel, "C," and kernel-specific hyperparameters like "gamma." HBO mimics the honey badger's ability to balance exploration and exploitation to find the best solutions.

The need for HBO-based parameter tuning arises from the fact that hyperparameters significantly impact SVM performance. Finding the right combination of hyperparameters can make the difference between a well-performing and a poorly performing model. HBO's ability to efficiently explore the hyperparameter space can lead to improved model accuracy and robustness. The pseudocode for the proposed multiplier construction is given below:

```
# Define the No. of honey_badgers
population_size = N
# Initialize a population of honey badgers with
random hyperparameters
population = initialize_population(N)
# Define the objective function to be optimized
(e.g., classification accuracy)
def objective_function(hyperparameters):
  # Train an SVM model with the given
hyperparameters
    model = train_svm_model(hyperparameters)
        # Evaluate the defined model's performance
    accuracy = evaluate_model(model, validation_data)
          return accuracy
# Set the maximum number of iterations for the HBO
algorithm
```

```python
max_iterations = max_iter
# Set the initial exploration and exploitation
parameters
exploration_param = initial_exploration_param
exploitation_param = initial_exploitation_param
# Initialize the best fitness and corresponding
hyperparameters
best_fitness = 0 # Initialize with a low value
best_hyperparameters = None
# Main optimization loop
for iteration in range(max_iterations):
  # Calculate the fitness of each honey badger in the
population
    for i in range(population_size):
        fitness = objective_function(population[i].
hyperparameters)
        population[i].fitness = fitness
            # Update the best fitness and
corresponding hyperparameters
        if fitness > best_fitness:
            best_fitness = fitness
            best_hyperparameters = population[i].
hyperparameters
  # For each honey badger in the population
  for i in range(population_size):
    # Update exploration and exploitation
parameters based on the iteration number
        exploration_param = update_exploration_
param(iteration, max_iterations)
        exploitation_param = update_exploitation_
param(iteration, max_iterations)
                # Determine the direction of movement
(exploration or exploitation) for the honey badger
        if random() < exploration_param:
            # Explore: Generate a random
hyperparameter within valid bounds
            new_hyperparameters = generate_random_
hyperparameters()
        else:
            # Exploit: Modify the current
hyperparameters
            new_hyperparameters = exploit_
hyperparameters(population[i].hyperparameters,
exploitation_param)
```

The process begins by defining the population size, which determines the number of "honey badgers" in the population. These honey badgers represent sets of SVM hyperparameters. The population is initialized with random hyperparameter configurations, providing a diverse starting point for optimization. The objective function plays a central role. It measures the performance of the SVM model with a given set of hyperparameters. Typically, this could be a classification accuracy metric on a validation dataset. The objective function is essential because the HBO algorithm's objective is to maximize it. The better the model performs, the higher the fitness value assigned to the hyperparameters.

The heart of the algorithm lies in this main optimization loop. The loop runs for a predefined number of iterations (max_iterations). In each iteration, the algorithm evaluates the fitness of each honey badger in the population. Fitness is determined by applying the objective function to each set of hyperparameters, reflecting how well the SVM model performs with those hyperparameters. The algorithm maintains awareness of the best-performing set of hyperparameters. As fitness values are computed, the algorithm compares each honey badger's fitness to the best fitness observed. If a honey badger's hyperparameters yield better results, the best fitness is updated, and the corresponding set of hyperparameters is recorded as the best solution found.

The HBO algorithm distinguishes between exploration and exploitation phases controlled by the coefficient factors. These parameters are altered during every iteration in order to balance the exploration and exploitation ranges. In each iteration the algorithm decides whether to explore or exploit each honey badger. If exploration is favored, the algorithm generates random hyperparameters within valid bounds.

In contrast, during exploitation, it modifies the current hyperparameters. This choice is guided by the exploration and exploitation parameters. The generated or modified hyperparameters are checked to ensure they fall within valid bounds. For instance, in the case of SVM hyperparameters, the regularization parameter (C) should be positive, and the kernel-specific parameters should also adhere to appropriate constraints. If the newly generated or modified hyperparameters improve the fitness of a honey badger, the hyperparameters are updated for that honey badger. This process mimics the honey badger's ability to adapt and refine its approach as it searches for prey. After the optimization process concludes, the algorithm returns the set of hyperparameters associated with the honey badger that achieved the best fitness. These hyperparameters represent the optimal configuration for the SVM model, which should lead to the best performance on your specific problem.

7.5 EXPERIMENTAL RESULTS

The results of our proposed system are implemented in Xilinx 12.1 using Verilog. This programming is used to compare the resources between existing and proposed systems. The proposed model is compared in terms of the slice, logic tables, and delay. When comparing the slices, the proposed model shows a smaller slice than the other model. The proposed system demonstrated a marked reduction in logic elements employed and shows the more reasonable utilization of area in the FPGA board. The "LUT" metric is used to evaluate the logical utilization of the proposed model

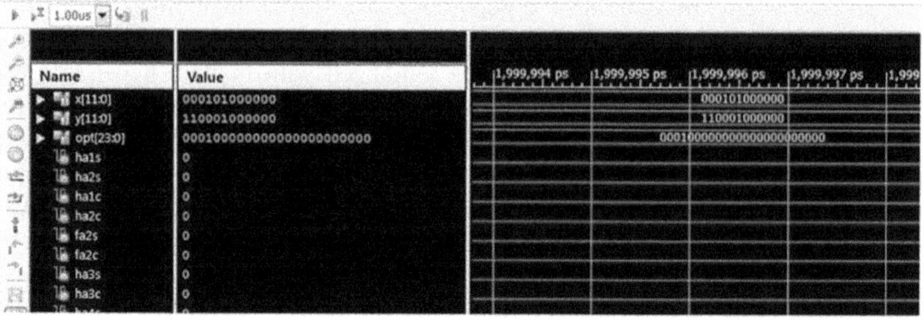

FIGURE 7.2 Product output.

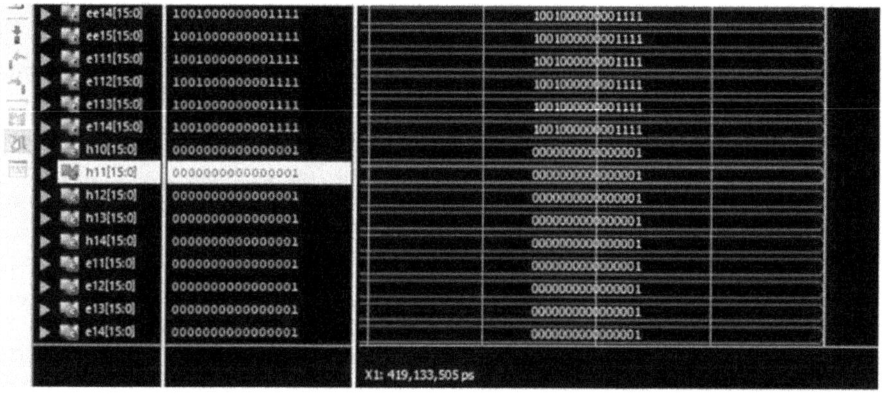

FIGURE 7.3 Final output.

with other models. These findings underscore the positive impact of our approach, which leverages SVM integration in FPGA design. The simulated waveforms are given in Figures 7.2 and 7.3.

The observed resource efficiency of SVM is very high when compared to existing systems. These results prove that more streamlined utilization of logic elements and combinatorial logic resources. The reduction in resource consumption can be attributed to the optimized precision of multipliers and careful compensation strategies. This outcome underscores the suitability of our approach for resource-constrained environments where the efficient allocation of hardware resources is paramount.

Moreover, our system's resource efficiency not only conserves FPGA resources but also holds the potential for enhanced power efficiency. As FPGA devices increasingly find applications in edge computing and IoT contexts, the system's ability to balance computational speed and precision holds particular relevance.

The performance of the proposed system is given in Table 7.1. Slice is a metric in FPGA terminology that refers to a group of configurable logic elements. A lower number of slices in the "With SVM" configuration compared to "Conventional" and "Without SVM" suggests that the implementation of SVM is more resource-efficient in

Device Utilization Summary(estimated values)			[-]
Logic Utilization	Used	Available	Utilization
Number of Slices	1219	960	126%
Number of Slice Flip Flops	360	1920	18%
Number of 4 Input LUTs	2095	1920	109%
Number of bonded IOBs	33	66	50%
Number of GCLKs	6	24	25%

FIGURE 7.4 Existing multiplier summary.

Design Goal:	Balanced	Routing Results	
Design Stratergy:	Xilinx Default(unlocked)	Timing Constraints	
Environment:	System Settings	Final Timing Score	

Device Utilization Summary(estimated values)				[-]
Logic Utilization	Used	Available	Utilization	
Number of Slices	612	960		63%
Number of 4 Input LUTs	1106	1920		57%
Number of bonded IOBs	33	66		50%

Detailed Reports					[-]
Report Name	Status	Generated	Errors	Warnings	Infos
Synthesis Report	Current	Wed Nov 23 14:47:46 2022		0 24 Warnings(21 new)	0
Translation Report					
Map Report					
Place and Rout Report					
Power Report					

FIGURE 7.5 Proposed multiplier summary.

TABLE 7.1
Performance Analysis

S. No	Parameter	Conventional	Without SVM	With SVM
1	Slice	1219	612	505
2	LUT	2095	1106	856

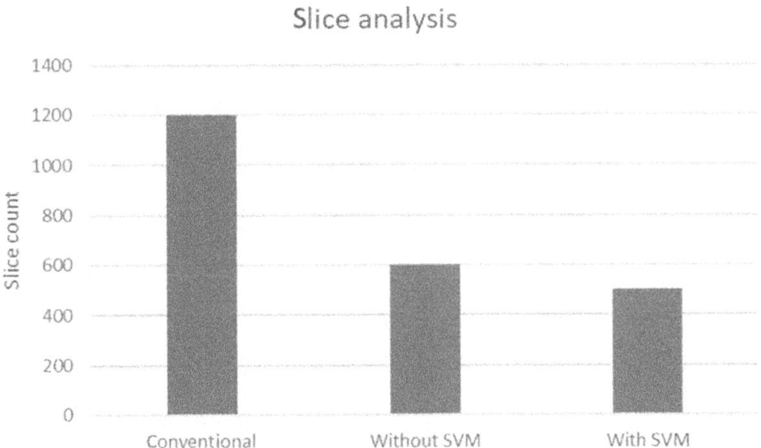

FIGURE 7.6   Area analysis.

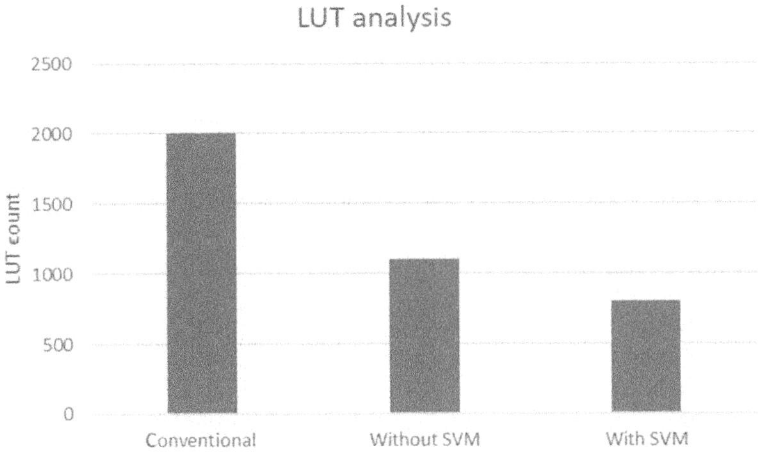

FIGURE 7.7   Delay analysis.

TABLE 7.2
Power Analysis

S. No.	Parameter	Conventional	Without SVM	With SVM
1	Power (nW)	12	9.5	7.6

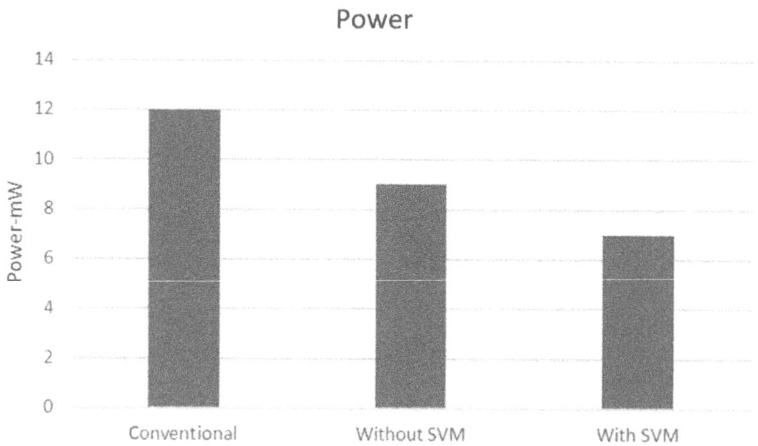

FIGURE 7.8   Power analysis.

terms of logic elements. Look-up tables (LUTs) are used for implementing combinatorial logic. The number of LUTs used in the "With SVM" configuration is lower than in both the "Conventional" and "Without SVM" configurations. This indicates that the SVM-based system is more resource-efficient in terms of logic resources, as shown in Figures 7.4 and 7.5. The comparison is graphically shown in Figures 7.6 and 7.7.

Table 7.2 shows the power analysis of the proposed model and it's clear that the proposed model "With SVM" consumes less power (7.6 nW) than the model "Without SVM" (9.5 nW). The power analysis is graphically shown in Figure 7.8.

7.6 CONCLUSION

In conclusion, the HBO algorithm presents a nature-inspired approach to the challenging task of tuning SVM hyperparameters. This optimization method leverages the behaviors of honey badgers in foraging for prey, balancing exploration, and exploitation to find the best set of hyperparameters for an SVM model.

REFERENCES

[1] M. Capra, B. Bussolino, A. Marchisio, M. Shafique, G. Masera, and M. Martina, "An updated survey of efficient hardware architectures for accelerating deep convolutional neural networks," *Future Internet*, vol. 12, no. 7, p. 113, Jul. 2020.

[2] F. Cardells-Tormo, P.-L. Molinet, J. Sempere-Agullo, L. Baldez, and M. Bautista-Palacios, "Area-efficient 2D shift-variant convolvers for FPGA-based digital image processing," in *Proc. Int. Conf. Field Program. Log. Appl.*, Tampere, Finland, 2005, pp. 578–581.

[3] L. Cavigelli, D. Gschwend, C. Mayer, S. Willi, B. Muheim, and L. Benini, "Origami: A convolutional network accelerator," in *Proc. 25th, Ed., Great Lakes Symp. (VLSI)*, New York, NY, USA, May 2015, pp. 199–204.

[4] S. Chakradhar, M. Sankaradas, V. Jakkula, and S. Cadambi, "A dynamically configurable coprocessor for convolutional neural networks," in *Proc. 37th Annu. Int. Symp. Comput. Archit. (ISCA)*, New York, NY, USA, 2010, pp. 247–257.

[5] D. Ali, A. U. Rehman, and F. H. Khan, "Hardware accelerators and accelerators for machine learning," *2022 International Conference on IT and Industrial Technologies (ICIT)*, Chiniot, Pakistan, 2022, pp. 01–07, DOI: 10.1109/ICIT56493.2022.9989124.

[6] Y. Braatz, D. S. Rieber, T. Soliman, and O. Bringmann, "Simulation-driven latency estimations for multi-core machine learning accelerators," *2023 IEEE 5th International Conference on Artificial Intelligence Circuits and Systems (AICAS)*, Hangzhou, China, 2023, pp. 1–5, DOI: 10.1109/AICAS57966.2023.10168589.

[7] S. C. Lee and T. H. Han, "A 4-way matrix multiply unit for high throughput machine learning accelerator," *2019 International SoC Design Conference (ISOCC)*, Jeju, Korea (South), 2019, pp. 113–114, DOI: 10.1109/ISOCC47750.2019.9078493.

[8] U. Gentile and L. Serio, "A machine-learning based methodology for performance analysis in particles accelerator facilities," *2017 European Conference on Electrical Engineering and Computer Science (EECS)*, Bern, Switzerland, 2017, pp. 90–95, DOI: 10.1109/EECS.2017.26.

[9] Minkwan Kee, Seung-jin Lee, Hyun-su Seon, Jongsung Lee, and Gi-Ho Park, "Intelligence boosting engine (IBE): A hardware accelerator for processing sensor fusion and machine learning algorithm for a sensor hub SoC," *2017 IEEE Symposium in Low-Power and High-Speed Chips (COOL CHIPS)*, Yokohama, Japan, 2017, pp. 1–3, DOI: 10.1109/CoolChips.2017.7946377.

[10] H. Chen and C. Hao, "Hardware/software co-design for machine learning accelerators," *2023 IEEE 31st Annual International Symposium on Field-Programmable Custom Computing Machines (FCCM)*, Marina Del Rey, CA, USA, 2023, pp. 233–235, DOI: 10.1109/FCCM57271.2023.00058.

[11] J. B. Song, Y. Kim, M. Lee, S.-S. Lee, and K. Kim, "A flexible machine vision ai system for edge-oriented deep learning accelerators," *2023 International Technical Conference on Circuits/Systems, Computers, and Communications (ITC-CSCC)*, Jeju, Republic of Korea, 2023, pp. 1–4, DOI: 10.1109/ITC-CSCC58803.2023.10212811.

[12] S. Brennsteiner, T. Arslan, J. S. Thompson, and A. McCormick, "A machine learning enhanced approximate message passing massive MIMO accelerator," *2022 IEEE 4th International Conference on Artificial Intelligence Circuits and Systems (AICAS)*, Incheon, Republic of Korea, 2022, pp. 443–446, DOI: 10.1109/AICAS54282.2022.9869942.

[13] B. Dong, Z. Wang, W. Chen, C. Chen, Y. Yang, and Z. Yu, "OR-ML: enhancing reliability for machine learning accelerator with opportunistic redundancy," *2021 Design, Automation & Test in Europe Conference & Exhibition (DATE)*, Grenoble, France, 2021, pp. 739–742, DOI: 10.23919/DATE51398.2021.9474016.

[14] Y.-K. Lai, K.-P. Chang, X.-W. Ku, and H.-L. Hua, "A machine learning accelerator for DDoS attack detection and classification on FPGA," *2022 19th International SoC Design Conference (ISOCC)*, Gangneung-si, Republic of Korea, 2022, pp. 181–182, DOI: 10.1109/ISOCC56007.2022.10031506.

[15] M. Zheng, F. Chen, L. Jiang, and Q. Lou, "PriML: An electro-optical accelerator for private machine learning on encrypted data," *2023 24th International Symposium on Quality Electronic Design (ISQED)*, San Francisco, CA, USA, 2023, pp. 1–7, DOI: 10.1109/ISQED57927.2023.10129302.

[16] N. Sutisna, Z. N. Arifuzzaki, I. Syafalni, R. Mulyawan, and T. Adiono, "Architecture design of Q-learning accelerator for intelligent traffic control system," *2022 International Symposium on Electronics and Smart Devices (ISESD)*, Bandung, Indonesia, 2022, pp. 1–6, DOI: 10.1109/ISESD56103.2022.9980698.

[17] C. Wang, L. Gong, X. Li, and X. Zhou, "A ubiquitous machine learning accelerator with automatic parallelization on FPGA," *IEEE Transactions on Parallel and Distributed Systems*, vol. 31, no. 10, pp. 2346–2359, 1 Oct. 2020, DOI: 10.1109/TPDS.2020.2990924.

[18] D. Aboye et al., "PyRTLMatrix: An object-oriented hardware design pattern for prototyping ML accelerators," *2019 2nd Workshop on Energy Efficient Machine Learning and Cognitive Computing for Embedded Applications (EMC2)*, Washington, DC, USA, 2019, pp. 36–40, DOI: 10.1109/EMC249363.2019.00015.

[19] V. Vranjković and R. Struharik, "Coarse-grained reconfigurable hardware accelerator of machine learning classifiers," *2016 International Conference on Systems, Signals and Image Processing (IWSSIP)*, Bratislava, Slovakia, 2016, pp. 1–5, DOI: 10.1109/IWSSIP.2016.7502737.

[20] Suresh, B. N. Reddy and C. Renu Madhavi, "Hardware accelerators for edge enabled machine learning," *2020 IEEE REGION 10 Conference (TENCON)*, Osaka, Japan, 2020, pp. 409–413, DOI: 10.1109/TENCON50793.2020. 9293918.

[21] S. Zhou et al., "ParaML: A polyvalent multicore accelerator for machine learning," *IEEE Transactions on Computer-Aided Design of Integrated Circuits and Systems*, vol. 39, no. 9, pp. 1764–1777, Sept. 2020, DOI: 10.1109/TCAD.2019.2927523.

[22] Z. Aizaz, K. Khare, and A. Tirmizi, "FASBM: FPGA-specific approximate sum-based booth multipliers for energy efficient hardware acceleration of image processing and machine learning applications," *2023 IEEE 31st Annual International Symposium on Field-Programmable Custom Computing Machines (FCCM)*, Marina Del Rey, CA, USA, 2023, pp. 210–210, DOI: 10.1109/FCCM57271.2023.00038.

[23] E. Nurvitadhi, A. Mishra, and D. Marr, "A sparse matrix vector multiply accelerator for support vector machine," *2015 International Conference on Compilers, Architecture and Synthesis for Embedded Systems (CASES)*, Amsterdam, Netherlands, 2015, pp. 109–116, DOI: 10.1109/CASES.2015.7324551.

[24] G. Xin, Y. Zhao, and J. Han, "A multi-layer parallel hardware architecture for homomorphic computation in machine learning," *2021 IEEE International Symposium on Circuits and Systems (ISCAS)*, Daegu, Korea, 2021, pp. 1–5, DOI: 10.1109/ISCAS51556.2021.9401623.

8 Realization of Smart City Based on IoT and AI

*Pavithra N., Sapna R., Preethi,
Manasa C.M., and Raghavendra M. Devadas*

8.1 INTRODUCTION

The term "smart city" is quite popular since it improves urban residents' quality of life by integrating many fields, like smart communities, smart transit, smart healthcare, smart parking, and many more. Real-time data processing and the ability to generate smarter decisions pose a substantial threat to the continuous evolution of complex metropolitan networks. The bounds of conventional networks have been widened by the phenomenal expansion of the many devices linked to the network. This significant development introduced the third wave of the internet, followed by social networking and static pages web (WWW), "Internet of Things (IoT)" [1]. Internet of Technology, prominent field which can identify as well as share data across heterogeneous objects that may be uniquely addressed. IoT has attracted interest from a variety of groups because of the expansion of embedded systems and the hasty growth in the number of devices being used. The Internet of Things (IoT) concept has expanded due to the attention of multiple interest groups and advancements in embedded device technology. This develops useful efforts, including smart housing, smart cities, smart health, etc. To effectively fulfill service requests, a huge volume of instantaneous data processing is required. General data processing and analytical procedures are unable to meet the demand for processing data instantaneously because of the enormous expansion in data volume. Therefore, it is thought that working with analysis is the best way to start building a smarter city. Academic and industrial professionals have made numerous attempts to realize the concept of intelligent cities. The present-day state of the art [2] documents numerous initiatives on particular areas of concern, such as water resource management, waste management, parking management, and many more. Big Data analytics is integrated with metropolitan IoT to realize the smart city vision. As an illustration, a smart meter placed in a home gathers readings from the meter and compares them to a predetermined power consumption limit. Based on the comparison, the smart grid is notified of the energy demand that exists at that moment. Consumers are simultaneously informed of the level of energy usage, enabling them to effectively manage energy use. Although the buzzword "smart city" has attained popularity in the era of technology, actual implementation is still in its early stages. Multiple initiatives are done in this regard to build a practical smart city.

DOI: 10.1201/9781032623276-8

The heterogeneous data will be stored and processed using Hadoop [3]. After Hadoop processing, intelligent decision-making linked to smart city functions is produced. The decisions' accompanying actions or events are then carried out. Numerous state administrations are embracing the urban communities and executing large information applications to arrive at the necessary degree of maintainability and work on the expectations for everyday comforts. Urban areas use various advances to work on the more elevated levels of solace to their residents. Shrewd urban areas include decreasing asset utilization and costs notwithstanding more effectively and successfully captivating their residents. One of the new advancements upgrades a huge amount of information in urban areas. The gathering of enormous volumes of data that can be used in useful fields has become a part of daily life. Success in numerous industries and sectors relies heavily on the utilization and effective analytics of huge quantity of data, commonly referred to as big data. Ninety percent of the world's digital data has been acquired in the previous 5 years alone, and big data growth is expected to exceed 40 percent every year [4]. As a result, numerous governments have started utilizing big data to support the growth and sustainability of smart cities. As a result, cities were able to maintain the standards, requirements, applications, and concepts of smart cities. Big data stores a lot of information, whether it is organized or not and whether it is structured. Big data is a typical database that comes through data processing. Large volumes of data will be efficiently stored by big data systems, which will then supply information to enhance the services that smart cities deliver. Using this information, big data will assist administrators in planning for any expansion in smart cities, locations, and resources. Many cities are transforming into smart cities because of advantages such as those for the environment, economy, and analysis. Therefore, we'll talk about some of the advantages and possibilities that could aid in transforming our city into a smart city. Through these advantages and possibilities, we may redesign our city to be a smart city. We can reach higher standards of sustainability, governance, and resilience by making use of benefits and possibilities. We can enhance natural resource management and life quality by implementing intelligent infrastructure management. Figure 8.1 illustrates the broader list of applications based on smart cities using big data analytics.

The following are some advantages of having a smart city: Effective resource management: As resources become either limited or very expensive, they must be used more carefully and integrated into solutions. Systems involving technology like corporate resource planning with geospatial information systems will aid this process. Locating waste distribution locations with a monitoring system will be easier, which will also help with energy conservation and resource management. Applications for interconnection and data gathering in smart cities that work together to provide services and applications.

8.1.1 Greater Quality of Life

Smart cities will have enhanced lifestyles because of vast housing options, services, and employment opportunities. The outcome is better and faster transit, adequate knowledge to make decisions, and better living and working conditions at a location.

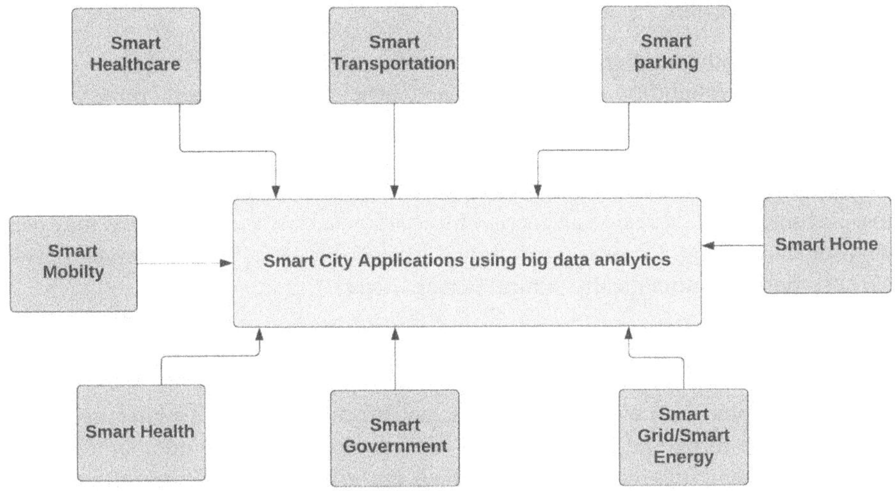

FIGURE 8.1 Overview of smart city application.

For these advantages to be realized, it is essential to establish high standards for the quality, privacy, control, and security of the data. It is also vital to employ data documentation standards to offer guidance on how to use content and data sets. This involves a high level of sophistication in the application as well as the involvement of people and resources.

8.1.2 SMART EDUCATION

A knowledge-based society will emerge, boosting capability and competitiveness. Big data and information and communication technologies will play a part in this. Big data in learning focuses on gathering information from students, teachers, parents, administrators, infrastructure like schools, library systems, educational locations, teachers, museums, and universities, as well as specialized data like tests, documents, finance, and assessments. It also involves collecting information from institutions such as universities. Data will develop into a valuable resource for information extraction and analysis to improve education. Data, for instance, aids learning for oneself when studying.

8.1.3 INTELLIGENT TRAFFIC LIGHTS

Issues with traffic, pollution, and the economy all develop with population growth. This is one of the most important strategies for managing heavy traffic and congestion in smart cities. Traffic grids about traffic patterns are coupled to smart signals and lights. Different aspects of traffic flow, such as traffic jams, vehicle speeds, and waiting times at traffic lights, are detected by sensors. These parameters are provided by the system, along with lights and signals.

8.1.4 Smart Grid

An essential element of smart cities. In order to gather data from electrical grid networks. The reliability, affordability, and efficiency of electric power are all improved. Smart remote controllers utilize two-way communication technology. The distribution, production, and transmission systems must have smart sensors and meters installed to get real-time data on current power usage and problems. It can give customers almost real-time energy information and let them buy what they need at reasonable pricing. Consumer electronics like washing machines and water heaters can cost more to automatically control during a sale.

8.2 CUTTING-EDGE TECHNOLOGIES

It is now simpler than ever to communicate information across platforms due to the IoT, which forms the foundation for sensors and devices to interact with each other in a smart-cities context. To enhance the effectiveness of smart cities, we may incorporate a smart city as a real system that connects many lives, including transportation and buildings. Due to the growing usage of many communication devices, IoT gadgets are positioned as the upcoming breakthrough technology by harnessing all the capabilities afforded by Internet technology. The development of city, like power grids, supermarkets, housing, water, transit, and buildings has recently become smart with the use of the IoT [5]. As part of the smart-city program, sensors and gadgets are integrated into hospitals, power grids, railways, pipelines, roads, houses, infrastructure, bridges, and other things all over the world. Storage facilities and strong computing capabilities are needed to support these types of smart applications. One method to provide such a platform is to use the various advantages of cloud services to enable data analytics in smart-cities management. Figure 8.2 illustrates the overall system architecture of a smart city using analytics of large volumes of data.

The massive amounts of data gathered from the current smart cities need to be precisely evaluated to create effective and universal smart city architecture. In a smart home or smart city ecosystem, sensors can be positioned in numerous areas to collect data. The ecosystem of an urban city is designed and planned, which can benefit from the offline processing of big data. However, it is useless for making decisions in the present. Numerous methods are built on the Hadoop ecosystem, where smart cities use big data systems to process, store, and mine data effectively. This helps to improve many smart-city applications tremendously. Big Data can indeed assist experts in designing and constructing smart-city services, assets, and infrastructure. Big data has a broad range of qualities that highlight its tremendous potential for growth and improvement.

The availability of cutting-edge technologies and techniques, however, places limits on the seemingly limitless possibilities. Huge volumes of data can attain their objective and improve processes in smart cities by utilizing proper methods and tools for effective and accurate analysis of data. Such efficiency would facilitate stakeholder cooperation and communication and make it easier to establish new facilities and technology that can improve the smart city. Big data applications could grant

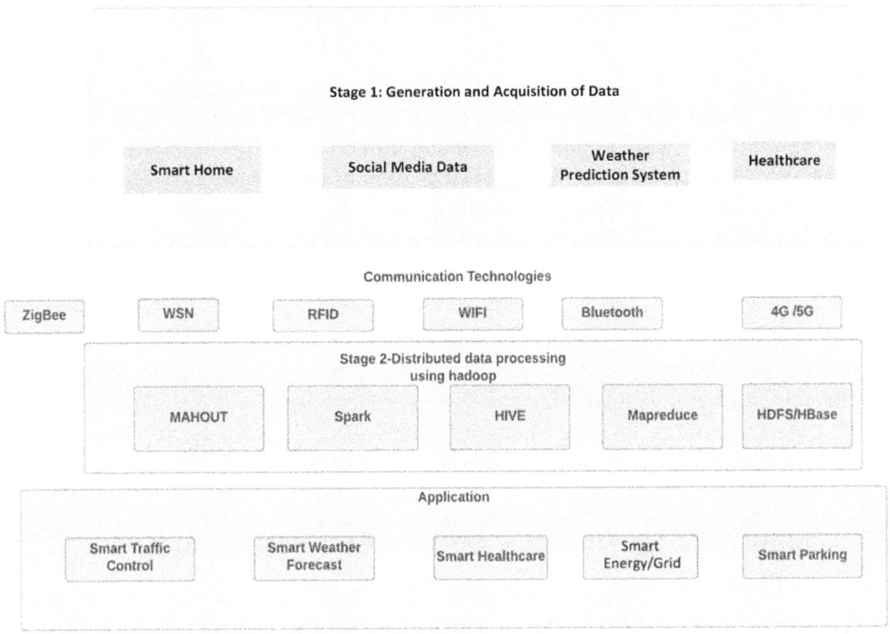

FIGURE 8.2   The system architecture.

many different areas of a smart city, enhancing customer experiences and services while also helping businesses operate more effectively. Healthcare can be enhanced by providing more patient attention, therapeutic and diagnostic instruments, management of medical records, and preventive care services. Big data has the potential to significantly increase environmental friendliness, adjust to fluctuating demand, and optimize routes and schedules for transportation networks. Big data is precisely what results from the production of numerous resources from databases. Smartphones, GPS, environmental sensors, computers, and even people's numerous apps, including games, advertising applications, digital photographs, social media sites, and greater numbers of faster data from previous times. Big data is a typical database that comes through data processing.

Big data is characterized by variety, value, velocity, and veracity. Large amounts of data will be efficiently stored by big data systems to supply knowledge that will enhance the services provided by smart cities. Big data will use this information to assist administrators in planning for any growth in smart locations, buildings, and resources. To provide a framework and coding tools for the parallel computing of huge datasets over numerous clusters, Hadoop was developed. MapReduce and Hadoop Distributed File System, the two main parts of Hadoop, are interconnected [6,7]. Despite the requirement for live/real-time data processing and storing in smart cities, the streaming style will provide good connectivity among various sensors inside the network. Such technology has lately become widely employed with the launch of different real-time data processing platforms, like Apache-Sturm, Apache-S4, Spark-streaming, and MLLib [7], which may permit data storage plus analysis via many networked nodes/clusters. "Cloud computing" refers to a variety of alternative computing architectures with several nodes/clusters linked by a real-time interaction system [8]. Large-scale, challenging, and time-consuming computer tasks, such as mining vast amounts of data from social networks generated by mobile applications, can be performed using cloud computing services [9]. IoT can be connected with cloud computing services like infrastructure-as-a-service (IaasS), software-as-a-service (SaaS), and platform-as-a-service (PaaS) [10]. This combination has the power to alter every business since big data technology makes it simple to analyze large data. Cloud computing can also offer a virtualized computing services platform that integrates security tools, data management tools, analysis software, visualization platforms, and customer deployment [10]. With the help of the cost-based paradigm that cloud computing may provide, users and organizations will be able to access data whenever they need it, from any location. Through the use of big data technology like the Hadoop framework, cloud computing also supplies the underlying engine.

8.3 EMERGENT TECHNOLOGIES FOR COMMUNICATION

Emerging technology for communication can include both highly visible examples, such as social media and smartphones, and less evident or hidden ones, such as speech recognition software or drones. Even though they are evolving quickly, the majority create a number of moral issues when made available to the general public. Emerging technologies are, by definition, novel, cutting-edge, and still under development, but they are anticipated to have significant socioeconomic effects. The user can typically communicate information or news through communication technology; however, it does not need to be immediate. That takes care of the essentials, but it's still a somewhat nebulous description that doesn't cover everything.

Smart cities are possible because of the smart networks that connect the user's gadgets, including cars, smart homes, and smartphones which are necessary for big data analytics. Data analytics should be able to be used by the smart network to receive and process data in an appropriate manner. Also, immediate responses should be sent to many organizations pertaining to smart cities. As the data grows in real-time applications, maintaining the quality of service (QoS) in smart networks becomes an important feature. Recently, distributed application events in these applications should be communicated in real time to the processing location. The events that follow may be sent straight from their sources, unfiltered and unaggregated, or directly as raw events [11].

IoT technology, which makes it possible for a wide range of objects to be detected and remotely controlled over current network infrastructure, must be put into practice if the smart city idea is to be adopted. This project will make opportunities for the adaptable integration of various smart city components. The addition of sensors and actuator components to the IoT will improve its accuracy and efficiency while also producing financial rewards.

An IoT technology will be the best option to accomplish this, as was previously stressed. Organizations like the Extensible Messaging Presence Protocol project, through its foundation like XMPP Standards Foundation, build a platform for IoT technology that is unconnected to any firm or cloud services. For instance, REST technology provides a scalable architecture that facilitates device interaction via the hypertext-transfer protocol (HTTP) and is tailored for IoT applications to allow communication between any device and a server [11].

RFID

Radio Frequency Identification [3] uses electromagnetic waves to recognize and track through tags affixed to things automatically. The tags, which come in active, passive, and battery-assisted passive varieties, retain information electronically. The battery-operated active kind periodically transfers its ID signal. The battery-free passive versions employ the radio-electric energy sent by the reader. With a small onboard battery, the passive battery-assisted model can only communicate when an RFID reader is nearby. RFID is a widely used embedded communication technology that may be used to identify almost everything, including people, clothing, pets, and even other people. Because of its wide range of applications, RFID is now perfect for smart cities; it may be used for cargo tracking, hospitals, and libraries.

The software manages and keeps track of the RFID tags that are added to user data. It is either a piece of unique software or a smartphone app. RFID software frequently includes a clever mobile app that functions in conjunction with it. This program can connect via Bluetooth or beacon technology with the RFID reader. With another electronic RFID-enabled gadget, communication is possible. The item with which you want to connect must have an RFID tag installed so that it can recognize the electromagnetic waves sent out by the RFID reader. The technology allows us to connect two RFID-enabled devices, one of which is an RFID reader and the other of which is an electronic gadget with an RFID tag. Some RFID readers are manufactured with built-in RFID antennas, whereas others include antenna ports. To use the readers, the users must add additional antennas [12].

Components of RFID used are as follows:

RFID tag: RFID tags come in passive and active varieties. A passive RFID tag is the barcode at the store. It doesn't require a power source, is easy to activate, and is paired with a certain object. Like the sensor tag behind the library book, an active RFID tag has a microchip that records data about the item and might also have an antenna or onboard sensor.

RFID scanner: A device that reads RFID tags and gathers evidence concerning the item or device to which the tag is attached is called an RFID scanner or reader. These tethered or portable readers support both USB and Bluetooth. Although not every barcode scanner has the capability to interpret an RFID tag, all RFID readers possess the ability to read a barcode. We can classify the RFID readers into three groups based on the range they cover, and they are described below.

- **Low-Frequency (LF) RFID:** When put within the range of 30 to 300 kHz, RFID-enabled devices can communicate with low-frequency RFID readers. They travel a lot less distance.

- **High-Frequency (HF) RFID:** High-Frequency RFID scanners are able to communicate with RFID-capable gadgets that operate between 3 and 30 Mhz.
- **Ultra-High-Frequency (UHF)-RFID:** The strongest RFID readers are ultra-high frequency (UHF) scanners because they can operate over a significantly wider range (up to 300 MHz).

RFID Software

This manages and keeps track of the RFID tags that are attached to the user's assets. It is only run by a piece of proprietary software or a mobile app. RFID software frequently includes a mobile app that functions in conjunction with it. This can communicate with the reader using Bluetooth or beacon technologies. Unlike barcodes, which are limited to certain products, RFID tags are universally compatible and their signal may be detected by a scanner even when they are covered.

Due to its potential, RFID is currently setting the standard for the creation of integrated gadgets that are fully trackable. For example, a meter can become a smart measuring device when an RFID tag is installed. RFID tags are used by many companies. For instance, the development of an automobile's assembly line while it is being produced can be observed. Animals can be positively identified thanks to the RFID microchips installed in livestock and companion animals. Through RFID-embedded technology, such an identification approach makes intelligent equipment for smart cities possible.

Wireless Sensor Networks (WSNs): It is a collection of wireless communication technologies and distributed autonomous sensing nodes that use low-power integrated circuits to disseminate data within the linked sensors. Various inexpensive, low-power, small devices that are coupled with one or more sensors are supported by this network. The sensor includes a radio transceiver for transmitting and receiving signals, a microprocessor, an electronic circuit for interacting with the sensing devices, and an energy source. The prospect of employing a sensor network with many intelligent sensors has increased because of the WSN's capacity to connect low-cost and small devices [13]. This network makes it possible to make smooth connections and share information between many contexts, which is efficient.

This feature of WSN makes it applicable to numerous fields, including water quality monitoring, machine health monitoring, and industrial process monitoring and control. The WSN is suitable for smart city integration since it can handle large-scale deployment in any setting [14]. The network provides a quick, affordable way to set up scattered monitoring and control equipment, avoiding the possible overhead expenses of wired systems. Real-time monitoring of physical and environmental factors including temperature, pressure, light, and humidity is possible with the WSN. These ambient variables are controlled by devices, like switches, motors, or actuators, using an effective wireless connection. These features make WSN meant for smart buildings, smart homes, and smart health. Despite the difficulties that WSN encounters regarding energy use.

Ultra-wideband, Zig Bee, and Bluetooth, Wireless Fidelity (WiFi) [16]: If a city has integrated wireless communication platforms, it might be referred to as smart.

A rapidly expanding technology, wireless communication offers rising levels of flexibility and mobility. Dynamic network development, low cost, and simple implementation are offered by wireless technology. Users can substitute traditional cable networks and access the Internet at broadband speeds when they are linked to an access point or using WiFi in ad hoc mode. Ultra-wideband technology focuses on high-bandwidth indoor short-range wireless networks with multimedia connectivity. ZigBee is designed for short-range wireless communication and has the option of a long battery life [15].

The goal of the Bluetooth standard is to replace bulky, expensive cords with low-cost, short-range devices for computer input-output devices like printers, mouse, keyboards, and any others. These short-range wireless technologies, which require a low-power network, have been essential for wireless data transmission. However, the QoS offered by wireless communication technology ought to improve because specific electronic devices have a strong leaning to travel from one network to another without encountering communication problems [16]. The creation of self-healing, self-organizing wireless communication networks is crucial to the success of a smart community.

5G, Long-Term Evolution Advanced (LTE-A), 4G Long-Term Evolution (LTE) and 5G [17]: The 4G wireless network, a development of the current 3G wireless standards, is called LTE technology. By extending this concept, 4G transforms hybrid data and voice networks into data-only IP networks. Additionally, 4G combines orthogonal frequency division multiplex (OFDM) and multiple input multiple output (MIMO) to increase data throughput over 3G [17]. A transmission method called OFDM makes use of several widely spaced carriers that are modulated at low data rates. This method uses a spectral efficiency system that allows for large data rates and the sharing of a single channel by several users. MIMO employs several antennas at the transmitter and receiver to significantly increase data throughput and spectral efficiency. It is anticipated that most of the machine-to-machine communication traffic will use 4G wireless networks.

By introducing high bandwidths, LTE-Advanced (LTE-A) enables the transition from fourth generation to fifth generation. It promises speeds that are nearly three times as fast as those of the basic LTE network, as well as includes features like carrier aggregation, more MIMO, coordinated multipoint, relay stations, and diverse networks. Furthermore, A more advanced technology known as 5G supports a bandwidth of up to 10 Gigabit per second with a comparatively low delay and gives way for connecting 100 billion plus devices. These technologies are still in their infancy and are presently being tested in a number of pilot projects. It is anticipated that 5G networks will be widely deployed by 2020. Fast and reliable Internet access as well as support in order to construct smart cities will be made possible by the introduction of 5G networks. New designs, including cloud RAN and virtual RAN, will be introduced by these networks, which can permit most server farms via localized data centers at the network edges, and increasingly centralized network setup [17].

Network function virtualization: Network service providers have been worried about managing network infrastructure. This issue hampers innovation in the telecom sector and impacts income, too. In order to reduce or completely eliminate their reliance on proprietary hardware, network operators. Network function virtualization emerged from the need to change the paradigm away from relying solely on hardware to perform key network functions (Network Functions Virtualization [NFV]) [18]. Industrial NFV specifications were created by a collaboration of telecom carriers from the European Telecommunications Standards Institute. NFV is a technique developed to benefit from the development of virtualization in information technology. Utilizing commercially available hardware, this technology converts network functions that are hardware-based into software-based applications.

A few benefits of this emerging technology include platform transparency, expansion, adaptability, enhancements in efficiency, lower capital, and operational expenses. The following elements are supported by NFV: physical servers, hypervisors, and guest virtual machines. Physical resources like RAM, storage space, and CPU are offered by the physical server. The environment in which a guest virtual machine can function is created by a virtual machine monitor known as a hypervisor [18]. The necessary software is started on the guest virtual machine, a piece of software that mimics the appearance and functionalities of an actual platform.

Quantum networking: In some respects, we were aware of a class of issues that a computer would never be able to handle even before the first actual computer. In 1994, Peter Shor accomplished a similar accomplishment by creating an algorithm for a yet-to-be-built quantum computer that can do prime factorization in a finite amount of time. Long before any such machine had been constructed, the potential (and constraints) of this unique sort of mechanical processing were being outlined. In the field of computing, quantum computers are a new, potentially revolutionary technology [19]. Quantum Networking, a related upcoming technology, allows quantum bits (qubits) to be transmitted between quantum networks. Like many others, there is a lack of any specific knowledge on whether quantum networking will be an obscure detour in the development of digital networks or whether it will establish the standard mainstream basis for tomorrow's digital services.

Quantum networking utilizes principles of quantum mechanics to enable secure and efficient communication and information processing. Quantum networking involves the development of technologies and protocols for the secure transmission of quantum information over long distances. It leverages quantum phenomena such as entanglement and superposition to achieve tasks such as quantum key distribution (QKD), quantum teleportation, and distributed quantum computing. Applications of quantum networking include quantum cryptography for secure communication, distributed quantum computing for solving complex problems, quantum sensor networks for high-precision measurements, and quantum-enhanced machine learning for accelerated data analysis.

8.4　BIG DATA IMPLEMENTATIONS IN IOT-ENABLED CITY OF THE FUTURE

Cities all across the world have adopted the buzzword "smart city," which is gradually emerging as a dominant concept for urbanization. Big data technologies are used in smart cities to populate data and enhance many solutions for smart urban areas. Big data can be used to build any resources or services needed for a smart city [20]. Big Data requires the appropriate concepts for effective data analysis in order to attain its objectives and enhance utilities in smart urban areas. These methods and tools may promote interaction and collaboration among entities, offer assistance to many industries in the smart city, and enhance customer perspectives and investment prospects.

- **Smart power grid**

This idea is inextricably linked to the emergence of sustainable sources that are ever-evolving and reducing dependency on fossil fuel substitutes while maintaining equilibrium between supply and demand [20]. A typical component to spot in this is a dual-directional energy flow, which suggests that users are no longer only passive consumers but also energy producers [21]. The main challenge for the smart grid is the networks themselves, as the majority of them were built to serve massive plants that run on fossil fuels, supplying power to customers only in a single way [22]. Three essential elements are required for this bidirectional flow: storage technology, management systems, and infrastructure supervision. Technical assistance is encouraged by each of these elements.

Studies have been trying to incorporate, assess, and use actual information on electricity generation and demand, as well as the types of climatic data due to this rapid spread of smart grids. Significant investment effectiveness of the current smart grid infrastructure is anticipated as a result of the improved system performance and various performance metrics. A significant amount of data is produced in this environment by varied sources like user energy consumption patterns, phasor measurement data for insight, and power usage data measured by widely used smart meters between multiple other sources [23].

Big data gathered from the smart grid environment used effectively will aid to take decision in making wise decisions regarding the quantity of electricity to be offered along with meeting user demand. Moreover, the analysis of the smart grid data can help to achieve well-planned objectives by providing a particular assessment plan that is reliable as per the source, request, and functional prototypes. The data analysis of these data could be used to foresee the future demand for power supply [24].

- **Smart healthcare**

The concept of "Smart Planet," first presented by International Business Machines Corporation (IBM), sparked the creation of smart healthcare. This concept is an intelligent system that collects data from sensing devices, transfers it over the IoT, and then uses cloud computing and processing power to process it [25]. In terms of managing global civilization in an interactive manner, it can synchronize and incorporate social systems. "Smart healthcare" refers to healthcare that makes use of the latest trends like wearables, the IoT, and mobile internet to gather data vigorously, interlink various healthcare-related components and organizations, and then vigorously accomplish as well as intelligently react to the requests of the health eco-system.

IoT-enabled medical care has the potential to facilitate communication among all involved parties in the healthcare sector, guarantee that participating individuals obtain the services they need, support decision-making, and encourage the effective use of technology. Smart healthcare is about raising the bar for information creation within the medical system.

The medical field has produced a vast amount of material over the years. The world's population increase has enabled abrupt changes in diagnosis deployment methods, with various conclusions driving these changes being data-driven. Healthcare professionals possess the ability to collect and assess medical data that can help administrative organizations and health insurers by using the appropriate analytics technologies. Also, efficient analytics of large medical data sets can improve quality of life, prevent unnecessary deaths, and help predict the emergence and spread of treatment options and diseases. Smart devices that are linked to homes or clinics to keep a check on patients, might raise the volume and consistency of information collected for certain individual's health conditions. This information is used to interpret clinical notes. Additionally, the analysis of vast volumes of healthcare data enables physicians to recognize warning symptoms of a severe illness during the initial therapy phases, potentially saving many patients [26].

- **Smart Transportation**

The growth of smart transportation can be impacted by transportation governance. Smart transportation governance in smart cities consists of a number of plans, regulations, strategies, and projects. Automated ticketing, electric automobiles, automated vehicles, and clean transportation legislation are a few examples of these approaches. In the transportation industry, the term "smart" can refer to new forms of propulsion (such as electricity), improved vehicle controls (such as the Intelligent Transport System), improved business models (such as car sharing), improved regulations, and improved transport planning and policy. Their primary goals are to reduce pollution, transportation congestion, increase safety, accelerate transfers, and lower travel expenses [27].

In order to reduce traffic congestion by offering alternate routes and lowering accident rates by evaluating the mistakes in the past, along with aspects like finding the reason for the mistake and driving speed, the patterns discovered out of the immense quantities of traffic data can aid the system of commuting. Transport systems' data can also be used to optimize freight movements [28]. The large amounts of data collected by smart transportation systems can also help with cargo optimization and consolidation by reducing supply chain waste. The data gathered from the IoT-enabled transportation system can also offer various advantages, including lowering environmental impact, boosting safety, and enhancing end-to-end customer experience.

- **Smart Governance**

Governments are having to reevaluate their obligations in a knowledge-based society as a result of the expanding role that technologies play in the operation of municipal systems. Prior studies have called this capacity "smart governance" [29].

Despite its significance, there isn't a consensus on what this idea means. Many have concentrated on the procedure for gathering various types of data as well as info pertaining to public administration by sensors or sensor networks [29] or on the accomplishment of the social integration of city dwellers using government services, whereas some earlier research focuses on both political engagement and the government's operation [30].

Smart governance can be enabled by big data analytics [31]. Analysis of data that can lead to collaborations between them can quickly identify organizations or agencies with similar goals. Collaboration like this can help countries develop. Since governments are well-versed in the requirements of individuals concerning healthcare and societal well-being, training, and other sectors, big data analytics may also help them create and execute effective policies. A review of comprehensive data from multiple academic institutions may also help to reduce unemployment.

8.5 TECHNICAL CHALLENGES

The increasing need for big data and smart cities fosters innovation and highlights the importance of developing new smart apps. However, good data management is required in order to enhance the smart city's services. These technological issues with big data and smart cities are briefed in this subsection. It takes the integration of several technologies to create a smart city. The adoption of smart city technology is further hampered by the QoS given by various technologies [32]. For instance, the aim to fulfill core objectives like fault-tolerant, scalable, and dependable networks cannot be sacrificed. Similarly, it is challenging to design massively scalable data analysis and storage systems supported by wise cloud service choices. Before the urban planning application is fully integrated, the QoS offered by these technologies must be adequate. The approaches and frameworks for establishing and implementing QoS criteria in a smart city are crucial [33]. Artificial intelligence is a subset of knowledge engineering that makes use of computational intelligence concepts like neural networks, genetic algorithms, and bioinformatics. These algorithms are effective, efficient, and robust [34]. However, small data sets are the only size for which machine learning algorithms are effective, efficient, and robust [35]. As a consequence, data analytics for smart cities cannot use these algorithms. Because smart cities generate large volumes of data, conventional computational intelligence approaches are no longer applicable to large data analytics.

With the aid of a variety of embedded intelligent units, data from smart cities may be gathered in many different ways. Data aggregation among these cities represents one of the greatest difficulties that must be overcome, though, as the target of the smart-city drive is to mix massive data generated within various origins [36]. Smart cities have included a number of technologies in recent years, lowering the technological obstacles to handling the data. However, one of the toughest issues with any data combining system is quality data, particularly if the information is unreliable, incomplete, inconsistent format, or insufficient [37]. Data analysis is regarded as an essential tactic for fostering progress and welfare in every modern city. To enhance the life quality for inhabitants & create reliable cities, processing issues with this data must be resolved [38]. In a smart city, data is extracted from numerous sources. To interpret the data and make decisions, one needs special algorithms and visualization techniques, which have an impact on activities related to smart cities. For instance, synchronizing the consumption detected by consumers' meters with that estimated by other utilities' systems helps reduce energy/water losses due to defective systems [39]. In the big data era, personal data in smart cities is susceptible to analytics, trade, and misuse, which creates worries about control loss, theft, and profiling [39]. People-specific information about residents, such as social events and places, is gathered daily, for instance. The massive amounts of personal data that smart city technology has collected are being protected against theft and hacking despite countless attempts to do so. A number of cyber-security issues related to smart city technology need to be addressed, even though city-level successful cyberattacks are still comparatively infrequent.

8.6 CONCLUSION

This chapter delves into the burgeoning proliferation of interconnected devices within urban landscapes, capturing the interest of scholars across diverse disciplines. The central focus is a comprehensive exploration of the utilization of big data analysis within the context of smart cities. The genesis of the "smart cities" idea is intrinsically connected to the remarkable expansion of the IoT, catalyzing the evolution of urban infrastructures toward intelligence. Despite this paradigm shift, the maturation of smart cities is still in its nascent phases, necessitating continuous innovation, robust networking frameworks, and adept data management capabilities. Establishing a pragmatic foundation for smart cities holds significant esteem among both academic circles and industry practitioners. This analysis elucidates pivotal technologies integral to intelligent urban frameworks, offering insights into specific smart city initiatives that stand to benefit significantly from the application of big data analytics. Additionally, prospective business models and architectural intricacies are scrutinized, with a distinct emphasis on the efficacious handling of voluminous data sets. In summation, the discourse underscores the pivotal role of extensive data in facilitating informed decision-making and extracting valuable knowledge within the ambit of smart city development.

REFERENCES

[1] Nathali Silva, Bhagya, Murad Khan, et al. "Big data analytics embedded smart city architecture for performance enhancement through real-time data processing and decision-making". *Wireless Communications and Mobile Computing*, no. 1 (2017): 9429676.

[2] Susmitha, K., & S. Jayaprada. "Smart cities using big data analytics". *International Research Journal of Engineering and Technology (IRJET)* 4, no. 8 (2017): 1615–1620.

[3] Hashem, Ibrahim Abaker Targio, Victor Chang, et al. "The role of big data in smart city". *International Journal of Information Management* 36, no. 5 (2016): 748–758.

[4] Silva, Bhagya Nathali, Murad Khan, Changsu Jung, & Jihun Seo. "Urban planning and smart city decision management empowered by real-time data processing using big data analytics". *Sensors* 18, no. 9 (2018): 2994.

[5] Gubbi, J., Buyya, R., Marusic, S., & Palaniswami, M. "Internet of Things (IoT): A vision, architectural elements, and future directions". *Future Generation Computer Systems* 29, no. 7 (2013): 1645–1660.

[6] Hashem, I. A. T, Yaqoob, Anuar, N. B., Mokhtar, S., Gani, A., & Khan, S. U. "The rise of big data on cloud computing: review and open research issues". *Information Systems* 47 (2015): 98–115. http://dx.doi.org/10.1016/j.is.2014.07.006.

[7] Hashem, I. A. T., Anuar, N. B., Gani, A., Yaqoob, I., Xia, F., & Khan, S. U. "MapReduce: Review and open challenges". *Sciento Metrics*, (2016): 1–34. http://dx.doi.org/10.1007/s11192-016-1945-y.

[8] Mell, Peter. "The NIST Definition of Cloud Computing." *NIST Special Publication* (2011): 1–7.

[9] Chang, V., David, B., Wills, G., & De Roure, D. "A categorisation of cloud business models." In *CCGrid, 10th International Symposium on Cluster, Cloud and Grid Computing*. 2010.

[10] Armbrust, M., Fox, A., Griffith, R., Joseph, A. D., Katz, R., Konwinski, A., & Stoica, I. "A view of cloud computing". *Communications of the ACM* 53, no. 4 (2010): 50–58.

[11] Atzori, L., Iera, A., & Morabito, G. "The internet of things: A survey". *Computer Networks* 54, no. 15 (2010): 2787–2805.

[12] Kang, Y.-S., Park, I.-H., Rhee, J., & Lee, Y.-H., "MongoDB-Based repository design for IoT-generated RFID/sensor big data". *IEEE Sensors Journal* 16, no. 2 (2016): 485–497. http://dx.doi.org/10.1109/jsen.2015.2483499.

[13] Dargie, W. W., & Poellabauer, C., "*Fundamentals of Wireless Sensor Networks: Theory and Practice*". John Wiley & Sons (2010).

[14] Al Nuaimi, E., Al Neyadi, H., Mohamed, N., & Al-Jaroodi, J., "Applications of big data to smart cities". *Journal of Internet Services and Applications* 6, no. 1 (2015): 1–15.

[15] Oualhaj, O. A., Kobbane, A., Sabir, E., Ben-othman, J., & Erradi, M., "A ferry-assisted solution for forwarding function in Wireless Sensor Networks". *Pervasive and Mobile Computing* 22 (2015): 126–135. http://dx.doi.org/10.1016/j.pmcj.2015.05.003Owen.

[16] Lee, J.-S., Su, Y.-W., & Shen, C.-C., "A comparative study of wireless protocols: Bluetooth, UWB, ZigBee, and Wi-Fi". *Paper presented at the 33rd Annual Conference of the IEEE Industrial Electronics Society, 2007*. IECON (2007).

[17] Abdalla, I., & Venkatesan, S., "Remote subscription management of M2Mterminals in 4G cellular wireless networks". *Proceedings of the 37th Annual IEEE Conference on Local Computer Networks* (LCN 2012): 869–877.

[18] Hawilo, H., Shami, A., Mirahmadi, M., & Asal, R. "NFV: State of the art, challenges, and implementation in next generation mobile networks (vEPC)". *IEEE Network* 28, no. 6 (2014): 18–26.

[19] P. W. Shor, "Algorithms for quantum computation: discrete logarithms and factoring". *Proceedings 35th Annual Symposium on Foundations of Computer Science* (1994): 124–134. http://dx.doi.org/10.1109/SFCS.1994.365700.

[20] Crossley, P., & Beviz, A. "Smart energy systems: Transitioning renewables onto the grid". *Renewable Energy Focus* 11 (2010): 54–59.

[21] Lund, H., Andersen, A. N., Østergaard, P. A., Mathiesen, B.V., & Connolly, D. "From electricity smart grids to smart energy systems—A market operation based approach and understanding". *Energy* 42 (2012): 96–102.

[22] "Renewables Won't Save Us If the Electric Grid Is Not Ready". Available online: www.forbes.com/sites/davidblackmon/ 2020/09/30/renewables-wont-save-us-if- the-electric-grid-is-not-ready/?sh=76da985c7abf (accessed May 23, 2022).

[23] Zhang, Y., Huang, T., & Bompard, E.F., "Big data analytics in smart grids: a review". *Energy Informatics*, 1, no 1 (2018): 1–24.

[24] Al Nuaimi, E., Al Neyadi, H., Mohamed, N., & Al-Jaroodi, J. "Applications of bigdata to smart cities". *Journal of Internet Services and Applications* 6, no. 1 (2015): 1–15.

[25] J. L. Martin, H. Varilly, J. Cohn, & G. R. Wightwick, "Preface: Technologies for a smarter planet". *IBM Journal of Research and Development*, 54, no. 4 (July–Aug, 2010): 1–2. http://dx.doi.org/10.1147/JRD.2010.2051498.

[26] Roy, N., Pallapa, G., & Das, S.K. "A middleware framework for ambiguous context mediation in smart healthcare application". *Third IEEE International Conference on Wireless and Mobile Computing, Networking and Communications, 2007. WiMOB* (2007).

[27] Benevolo, C., Dameri, R. P., & D'Auria, B. "Smart mobility in smart city". In: Torre, T., Braccini, A., Spinelli, R. (eds.) *Empowering Organizations. Lecture Notes in Information Systems and Organisation*, 11. Cham: Springer (2016). https://doi.org/ 10.1007/978-3- 319-23784-8_2.

[28] Ju, G., Cheng, M., Xiao, M., Xu, J., Pan, K., Wang, X., & Shi, F. "Smart transportation between three phases through a stimulus-responsive functionally cooperating device". *Advanced Materials* 25, no. 21 (2013): 2915–2919.

[29] Giffinger, Rudolf, Christian Fertner, Hans Kramar, Robert Kalasek, Natasa Pichler-Milanovic, & Evert J. Meijers. "Smart cities. Ranking of European medium-sized cities. Final Report." (2007).

[30] Schuurman, D., Baccarne, B., De Marez, L., & Mechant, P. "Smart ideas for smart cities: Investigating crowdsourcing for generating and selecting ideas for ICT innovation in a city context". *Journal of Theoretical and Applied Electronic Commerce Research* 7, no. 3 (2012): 49–62.

[31] Caragliu, A., Del Bo, C., & Nijkamp, P. "Smart Cities in Europe". *Proceedings to the 3rd Central European Conference on Regional Science*, Kosice, Slovak Republic, 45–59 (2009).

[32] Bellavista, Paolo, Antonio Corradi, & Andrea Reale. "Quality-of-service in data center stream processing for smart city applications." In: Samee U. Khan, Albert Y. Zomaya, & Elias Houstis. Cham (eds.), *Handbook on Data Centers*, 1047–1076. Cham, Switzerland: Springer (2015).

[33] Jalali, R., El-Khatib, K., & McGregor, C. (2015). "Smart city architecture for community level services through the internet of things". *Paper Presented at the 2015 18th International Conference on Intelligence in Next Generation Networks(ICIN)*.

[34] Jin, J., Gubbi, J., Marusic, S., & Palaniswami, M. "An information framework for creating a smart city through internet of things". *IEEE Internet of Things Journal* 1, no. 2 (2014): 112–121.

[35] Tsai, C.-W., Lai, C.-F., & Vasilakos, A. V. "Future Internet of Things: open issues and challenges". *Wireless Networks* 20, no. 8 (2014): 2201–2217.

[36] Su, K., Li, J., & Fu, H. (2011). "Smart city and the applications". *International Conference on Electronics, Communications and Control (ICECC)* (2011).

[37] Gouveia, J. P., Seixas, J., & Giannakidis, G. "Smart city energy planning integrating data and tools". *Proceedings of the 25thInternational Conference Companion on World Wide Web* (2016).

[38] Gulisano, Vincenzo, Magnus Almgren, & Marina Papatriantafilou. "When smart cities meet big data." *Smart Cities* 1, no. 98 (2014): 40.

[39] Tene, O., & Polonetsky, J. "Privacy in the age of big data: A time for big decisions". *Stanford Law Review Online*, 64 (2012): 63.

9 Sentiment Analysis of Airline Tweets Using Machine Learning Algorithms and Regular Expression

S. Nagendra Prabhu, A.P. Rohith,
Shubhankar Bhope, and P. Sivakumar

9.1 INTRODUCTION

Sentiment analysis, as the name implies, is determining the perspective or feeling underlying an event. Essentially, it entails analyzing and determining the emotions and intents underlying text, speech, and modes of communication. We, humans, communicate in a variety of languages, but each language is merely an intermediary or means through which we seek to express ourselves. And we are attached to everything we say. It can be positive, negative, or neutral. Suppose you have a restaurant chain that sells a variety of foods such as dosa, paratha, naan, and milkshakes. You created a website to sell groceries. Clienteles may now order items from your website and even leave reviews.

User Review 1: I love this butter naan; it's so delicious.
User Review 2: This chicken dosa has a very bad taste.
User Review 3: I ordered this paratha today.

The first review was positive, and that means the customer was really satisfied with the butter naan. The second opinion is negative; therefore, the company must review its dosa division. The third does not mean whether the customer is satisfied, so we can take that as a neutral statement. However, now a problem arises: there will be hundreds and thousands of user reviews for their product, and after a while, reviewing each user review and making conclusions will become almost impossible. They also can't come to a conclusion by taking only about 100 reviews because maybe the first 100–200 customers tasted the same and liked the naan, but over time, they didn't. With rating increases, there can be a situation where none of the positive reviews can cross negative reviews.

DOI: 10.1201/9781032623276-9

FIGURE 9.1   Emoji for sentiment analysis.

9.2 TEXT SENTIMENT CLASSIFICATION

People love to express their feelings. Happy or unhappy. Like or dislike. Compliment or complaint. Good or bad, as shown in Figure 9.1. That is, positive or negative. Emotion analysis in natural language processing involves decoding these emotions from the text. Is it positive, negative, or both? If there is a sentiment, what object in the text does that sentiment refer to, and actual emotional phrases such as poor, lack-luster, cheap, etc. (not only positive or negative)? As a technique, sentiment analysis is both interesting and useful. First, the enjoyable part. It's not always clear, at least for a computer program, if a text's emotion is positive, negative, or mixed. The signals can be subtle. Common sense aside, it's even harder to discern which objects in the text are the subjects of which feelings, especially when both positive and negative feelings are involved. Emotion analysis focuses on a writing's polarities (good – positive, bad – negative, or neutral). However, it appears that divergence can distinguish individual thoughts and sentiments (angry, pleased, sad, etc.), urgency (essential, not important), and even purpose. You can create and modify classifications to meet your sentiment evaluation requirements, contingent on how you want to interpret consumer feedback and queries.

9.3 LITERATURE INVESTIGATION – GENERAL

This chapter affords an exhaustive explanation of the frequent processes and strategies we looked into and used to comprehend the approach necessary for our project. In order to maximize the speed, efficiency, and quality of classification and detection in our system, we improved our algorithm with the help of the reference papers we studied. These improvements allowed us to explore effective solutions to some of the problems that were present in the classification of text.

9.3.1 CURRENT SYSTEM

A literature study looks for systems that are comparable to existing ones and determines how the researcher's project differs from them. This makes it easier to understand the project in depth. It gives a subject-specific blend of theoretical, methodological, and current understanding of findings. Information must be gathered in unity with the project. The following part outlines our model and how it differs from other similar models.

A study by Schaffer and Huynh [1] conducted a pictorial proportional investigation of the various progressions to investigate the characteristics of the course assembly variations by enchanting into justification the students' behavioral characteristics. It produced several endorsements for instructors and course designers. In their research [2], Kevin Andrey and Liedy offered three methods for categorizing the moods of tweets. A strategy centered on word, emotion, and a hybrid approach. The sentiment infused into the emotions is taken into consideration as a criterion for categorizing the messages in the emotion-based method. The inclusion of an emoticon was a selection factor for the tweet. Words like bad, excellent, good, etc., that reflect sentiment are engaged in interpretation when using the word-based technique to infer sentiment. In the hybrid method, the use of words and emoticons was taken into consideration. Stoffova et al. [3] executed the three main experiments to determine the student's place of origin in relation to technical awareness in a global educational setting. Students from two nations on two distinct continents took part in this study. They used principal component analysis (PCA) and support vector machine (SVM) to do the same. Adarsh and Ravikumar [4] performed a similar study. By taking into consideration the tweets from three well-known Airlines, a performed efficient yet straightforward way of recognizing feelings on Twitter is proposed. Based on the score calculation, sentiments were classified as negative, neutral, or positive. Torre-Díez et al. [5] castoff two machine learning processes to perform the basic architecture. The process involves a data gathering phase, preliminary text processing phases over tumbling incorrect information, and expression-entrenching strategies for renovating text-based information into numerical-constructed information; Convolutional Neural Network's (CNN's) algorithm can automatically determine qualities. Feedforward Neural Network (FFNN) for manipulative neither positive sentiment score (PSS) nor negative sentiment score (NSS) requirements, and Mamdani Fuzzy System (MFS) for categorizing the data they provide as favorable, unfavorable, or neutral. Ishaq et al. [6] provide a sentiment evaluation categorization strategy that employs a CNN's besides a GN (genetic algorithm) strategy. After mining semantic structures, many techniques, including the recommended CNN-based ensemble, Support Vector Machine, random forests, maximum entropy, stable differential investigation, decision tree, and generalized linear conventional, are trained concurrently. Lei et al. [7] propose an algorithm for suggestions based on sentiment data extracted from user feedback on social media platforms. To achieve the assessment prediction goal, they combine user opinion, resemblance interaction sentiments influence, and element reputation similarity in a solitary matrix factoring method. They mainly rely on social media users' emotional responses to determine their preferences. Chen et al. [8] take an alternative strategy. A cross-modal hypergraph prototype was developed to combine documentary and emotive data for sentiment categorization, which is useful for producing online service predictions. Furthermore, the Latent Dirichlet Allocation (LDA) concept prototypes have been implemented into our projected cross-modal hypergraph framework, allowing us to better interpret global higher-level facts and eliminate ambiguity in certain specific phrases. Manshu and Bing presented a Hierarchical Attention Network with Priority (HANP) method for Context-Dependent Sentiment Classification (CDSC) employment [9]. When classifying sentiment, the recommended HANP can prioritize keywords and phrases. The sentiment vocabulary matching stage can identify key pivots and non-pivots. By learning synonyms, the HANP can identify and understand terms related to both pivot and non-pivot (or dis-pivot) concepts. Abdalgader and Shibli [10] presented a novel application of the lexicon-based keyword polarities detection approach to various customer evaluations based on the convention of contextual expansion. The modified methodology finds a word's polarities by examining the semantic relationships among its context enlargement, equivalents, and glosses.

Lee [11] studies the early effect of COVID-19 emotion on the United States stock market by consuming big data statistics, namely the day-to-day news emotion Index Daily New's Mood Index (DNSI) and Google's Tendencies information on coronavirus-connected search terms. The goalmouth is to explore links between COVID-19 emotions and 11 other US market segment directories over a certain period. Yu et al. [12] proposed a Chinese Emotional – Sentimentality evaluation approach that utilized a larger dictionary. The extended sentiment dictionary consists of normal sentiment dictionaries, field sentiment sentences, and polysemic field emotion keywords. The Naïve Bayesian field classifier determines the sentimentality polarity of a polysemic sentiment word within a text. Rathi et al. [13] suggested an article to discuss. To assess the classifier's overall precision in tweet categorization, they employed SVM, AdaBoosted Decision Tree, and Decision Tree constructed hybrid emotion cataloguing methods. Our suggested methodology classifies tweets as detrimental or favorable and then employs sentiment analysis to make subsequent decisions. We employ preprocessing techniques to offer accurate data as input for the training procedure. Zhong et al. [14] introduced stock-NF, a flow-oriented generating framework aimed at forecasting stock undertaking. To aid learning in being more adaptable and evocative, latent depictions of tweets besides antique stock values, the proposed method makes use of normalizing flows, which are extensively employed in the creation of pictures and audio synthesizing. Cheng and Yue [15] introduced Part-of-Speech and Targeted Attention Network (POS-TAN), a sentiment identification model they developed. The framework comprises the POS-attention mechanisms, which accumulate emotive content embedded in parts of speech, and the self-attention method, which is used to understand the text's feature expression. Furthermore, they employ the focal loss to mitigate the effects of sampling instability on the organization effect. Table 9.1 shows the literature survey of the sentiment analysis model.

9.4 LITERATURE REVIEW – GENERAL

- The most frequently used procedures for the classification of text are SVM, linear regression, and Naive Bayes.
- Multinomial NB is most effective for two-class issues, which presents a barrier when using it. This possibly will inhibit our effort because of the degree of the dataset.

The proposed method categorizes text based on sentiment using a Random Forest-based Classifier. The evaluation of a considerable number of research articles found that the following algorithms are most commonly utilized for sentiment analysis in text: support vector machine and random forest.

The challenge of using the support vector machine is that it is superlative suited for two-class complications. The size of the dataset may have an impact on our work.

TABLE 9.1
Literature Survey

Publication Details	Methodology	Algorithm Used	Advantages	Limitations
I. D. L. TorreDíez, et al., "Sentence Level Classification Using Parallel Fuzzy Deep Learning Classifier", *IEEE Access*, vol. 9, pp. 17943–17985, 2021.	The basic architectural hybrid model was carried out by I. D. L. Torre Dez, et al., data collection phase, the preprocessing OF texts processes to reduce noisy data, Methods for transforming text-based data into numeric representations using word embedding techniques, Convolutional Neural Network (CNN) methods for obtaining characteristics immediately, FFNN calculates PSS and NSS values, while MFS classifies the information it receives into neutral, negative, or positive classifications.	Deep learning classifier using parallel fuzzy.	Predicting highest accuracy 90% with this model.	• Time ingesting is more. • Space cast-off by a model is very large.
V. Stoffova, Z. illes et al., "Machine Learning-Based Student's Native Place Identification for Real-Time", *IEEE Access*, vol. 8, pp. 130840–130854, 2020.	The three main tests were carried [3] to determine the student's country of origin with regard to technical knowledge in a global educational setting. The participants in this research study were students from two nations located on opposite continents. They accomplished the same using PCA and SVM.	Support vector machine, PCA.	Clarity of data and ease of training.	• Limited number of extraction techniques. • Overfitting, data leakage.
Brandon Huynh and James Schaffer, "An Analysis of Student Behavior in Two Massive open online courses", *IEEE/AC M ASONAM*, 2016.	When making predictions, it seems to be conducted a pictorial proportional examination of the various progressions to examine the effects of the variations in development assembly while captivating into interpretation the behavioral characteristics of the students. They then provided a number of recommendations for teachers and course designers.	TF-IDF vector space models.	In this particular case study, employing a visual and systematic analytics approach proved effective in elucidating student achievement.	Course structure was considered when making predictions. The predictive power of the algorithm was not good.

(*continued*)

TABLE 9.1 (Continued)
Literature Survey

Publication Details	Methodology	Algorithm Used	Advantages	Limitations
T. Zhong, et al., "Learning Sentimental and Financial Signals With Normalizing Flows for Stock Movement Prediction", *IEEE Signals-Processing Letters*, vol. 29, pp. 414–418, 2022.	This article offered StockNF, a flow-based framework for forecasting fluctuations in stocks. The proposed methodology employs flow normalization to empower the development of more accommodating and communicative latent illustrations for tweets and previous stock values.	StockNF	Improved the f-measure of sentiment prediction and overall classification accuracy.	Overfitting, data leakage.
M. J. Adarsh and P. Ravikumar, "An Effective Method of Predicting the Polarity of Airline Tweets using sentimental Analysis", *2018 4th ICEES*, 2018.	By analyzing tweets from three well-known airlines, this article suggests an efficient yet straightforward method for identifying sentiment on Twitter. The calculation of the score served as the basis for identifying positive, negative, and neutral moods.	Score calculation Score = No. of positive words - No. of negative words.	We were easily able to find difference between positive and negative sentiment of twitter users.	The problem of this method is that it may not achieve the desired results when applied to mocking twitters since the placement of optimistic and undesirable results leads to differing conclusions.
Liedy's del Carmen et al., "Analysis of Behavioral of Customer in the Social Networks using Data Mining Techniques", *IEEE /ACM ASONAM*, 2016.	The methodology utilized was CRISP, which involved evaluating descriptive models through clustering and association rules. The findings indicate that the proposed models can offer valuable insights for crafting marketing strategies tailored to user preferences.	CRISP-data mining techniques.	This technology saves money since it employs a number of methods that eliminate easy data mining activities and are widely used in business.	Overfitting and increment in size of training datasets.

Lee, H. S. "Leveraging big data to investigate the initial effects of COVID19 mood on the United State stock market", *Sustainability*, vol. 12, no. 16, p. 6648, 2020.	Lee et al., examines the initial influence of COVID-19 sentiment on the US stock market by leveraging big data, specifically the Daily New's Mood Index (DNSI) and Google Trends data pertaining to coronavirus-related searches. The aim is to explore the correlation between COVID-19 sentiment and 11 distinct categories within the US stock market directories over a specified time frame.	Time series regression models.	The study identifies industry-specific COVID-19 attitudes and categorizes them based on correlation.	Overfitting
A. Ishaq et al., "Aspect Based Sentiment Analysis Using a Hybridized Approach Based on CNN and GA", *IEEE Access*, vol. 8, 2020.	A highly effective classification performance for sentiment analysis can be achieved using a CNN combined with a genetic algorithm. Additionally, Support Vector Machine, CNN-based ensemble methods, Random Forest, Entropy, Stabilized Discriminant Analysis, Decision Trees, and various other models are trained following the extraction of semantic data.	CNN and GA.	Parallelism, Global optimization.	Being stochastic, there are no guarantees on the optimality or the quality of the solution.
G. Zhao et al., "Rating Prediction Based on Social Sentiment From Textual Reviews," *IEEE Transactions on Multimedia*, vol. 18, no. 9, pp. 1910–1921, Sept. 2016.	This article proposes a recommendation mechanism based on sentiment data gathered from user evaluations on social broadcasting. To solve the rating prediction problem, they incorporate handler sentiment resemblance, interpersonal sentiment effect, and item character similarity in a solitary matrix factoring approach. They specifically exploit user sentiment on social media to demonstrate preferences.	RPS model, context-MF, PRM, and EFM.	The findings confirmation that the three sentimental components significantly influence the prediction of rating. On a real-world dataset, it also demonstrates notable gains over existing techniques.	False positive is very high.

(*continued*)

TABLE 9.1 (Continued)
Literature Survey

Publication Details	Methodology	Algorithm Used	Advantages	Limitations
T. Manshu et al., "Adding Prior Knowledge in Hierarchical Attention Neural Network for Cross Domain Sentiment Classification", *IEEE Access*, vol. 7, pp. 32578–32588, 2019.	They propose using the HANP technique to the CDSC problem in this chapter. When recognizing sentiment, the recommended HANP can focus on significant words and phrases. The emotion dictionary match layer tracks key pivots, non-pivots, and dis-pivots. By understanding its equivalents, the HANP can figure out what a non-pivots or dis-pivots is.	CNN and RNN.	The effectiveness of HANP has been tested using the Amazon review dataset. We achieve cutting-edge accuracy with this dataset.	Overfitting
Z. Chen et al., "TCMHG: Topic-Based Cross-Modal Hypergraph Learning for Online Service Recommendations", *IEEE Access*, vol. 6, pp. 24856–24865, 2018.	To enhance emotional categorization for online services, this article proposes the development of a cross-modal hypergraph model that integrates textual and emotional information separately. Additionally, the LDA (Latent Dirichlet Allocation) topic model is integrated into the proposed cross-modal hypergraph model to incorporate higher-level global knowledge and to reduce ambiguity in certain phrases.	Cross-modal hypergraph model, SVM, LDA topic model.	The theme-created mixture model can improve classification performance even further and cut down on computational expenses.	• High time complexity Overfitting

Reference	Summary	Technique	Findings	Drawbacks/Notes
A. Shibli et al., "Experimental Results on Customer Reviews Using Lexicon-Based Word Polarity Identification Method", *IEEE Access*, vol. 8, pp. 179955–179969, 2020.	This article presents a revised version of the lexicon-based word polarity identification approach that uses context expansion in response to several user critiques. The improved method computes the semantic relatedness of context expansions identified in WordNet synonyms and glosses in order to control the polarity of an estimated term.	Lexical analysis.	It was able to detect negation scope of the word.	• Less accuracy • 78% • Accuracy
A. Malik et al., "Sentiment Analysis of Tweets Using Machine Learning Approach", *2018 11th International Conference on Contemporary Computing (11th ICCC)*, pp. 1–3, 2018.	They suggest the POS-TAN sentiment categorization model in this chapter. This model comprises the POS attention, which is used to collect emotive information contained in parts of speech, in addition to the self-attention mechanism, which is used to learn the text's feature expression. In order to lessen the effect of sample imbalances on the classification effect, they also introduce focus loss.	SVM, Adaboosted Decision tree and decision tree.	Improved the F-measure of sentiment prediction and overall classification accuracy.	• Disregards the aspect-related data. A sentence may have numerous diverse features, and it is occasionally important to determine the sentiment polarity of a particular aspect.
F. Li et al., "Chinese Text Sentiment Analysis Based on Extended Sentiment Dictionary", *IEEE Access*, vol. 7, pp. 43749–43762, 2019.	In this article, a more sophisticated dictionary-based approach is proposed for analyzing Chinese sentiment. The extended mood dictionary combines the standard mood dictionary with additional field mood words and polysemic field mood words. The Naive Bayesian domain classifier is employed to classify the textual field, including polysemic emotional terms, enabling the determination of the emotion polarity associated with the words.	Naive Bayesian field classifier.	This approach guesses a test dataset's class accurately and quickly. It can be used to solve multiclass prediction problems since it works.	The Naive Bayes framework will assign zero probability to any identifiable variable in the experiment data set and will be unable to foresee anything in this regard if it is not present in the training dataset.

(continued)

TABLE 9.1 (Continued)
Literature Survey

Publication Details	Methodology	Algorithm Used	Advantages	Limitations
Y. Yue et al., "Sentiment Classification Based on Part-of Speech and Self Attention Mechanism", *Institute of Electrical and Electronics Engineers, Access*, vol. 8, 2020.	In this chapter, it deliver a approach that is used to increase the classifier's overall accuracy when classifying Twitter data. For the same reason, we employ preprocessing techniques to ensure that the training process receives valid data. Our proposed method categorizes tweets as good or negative, which improves sentiment analysis and can be used to inform future decisions.	SVM, Adaboosted decision Tree and decision tree.	Improved the F-measure of sentiment prediction and overall classification accuracy.	• Overfitting, data leakage.

Multinomial Naive Bayes is an alternative. The Naive Bayes procedure's primary purpose is to determine the possibilities of classes dispensed to texts using the combined possibilities of arguments and classes. We will also employ regular expressions to ensure our model accurately represents modern slang and emojis.

9.5 SYSTEM ANALYSIS

9.5.1 CONTEMPORARY SYSTEM LIMITATIONS

Some of the present systems use the following methods for the classification of text using sentiment: machine learning procedures similar to random forest, SVM, linear regression, etc. In traditional systems the algorithms are repetitive, like random forest, SVM, and linear regression. The difficulty we face while using SVM is that it does not implement very well when the data set is additional wide-ranging, i.e., target classes overlap.

9.5.2 PROPOSED SYSTEM

The suggested technique uses a machine learning methodology similar to linear regression and multinomial Naive Bayes with regular expressions together to progress the accurateness of the classical.

9.5.3 OBJECTIVE

Implementing regular expression and machine learning together will help progress the accuracy of the classical. Here, we create a list of positive and negative sentiments using regular expressions. This list includes all the modern text abbreviations, slang, and emojis. We will create a flag for each sentiment by looking at all patterns defined in the list. After the above step, we get a new dataset, and after applying the ML algorithm to the new dataset, we get better accuracy. Also, a comparative study will be conducted to display the actual difference in the classification when a machine learning system is used and when the machine learning system with regular expression is used.

9.5.4 PROPOSED ARCHITECTURE

The flow of architecture is explained in Figures 9.2 and 9.3, where we begin after taking the dataset of airline review tweets. Preprocessing and morphological feature extraction are performed on this dataset. We try to remove the noise and unwanted data from the dataset. Next, we will use Porter stemming for keywords in the processed datasets. We use Inverse Frequency-Targeted Document Frequency to check the position or relevance of string representations (words, expressions, lemmas, etc.) in a dataset or data. Next, it will practice regular expression to make an incline of negative and positive seeming words and emojis. In the next step, we will try to flag our dataset tweets depending on their positive or negative attributes. Next, it will train our

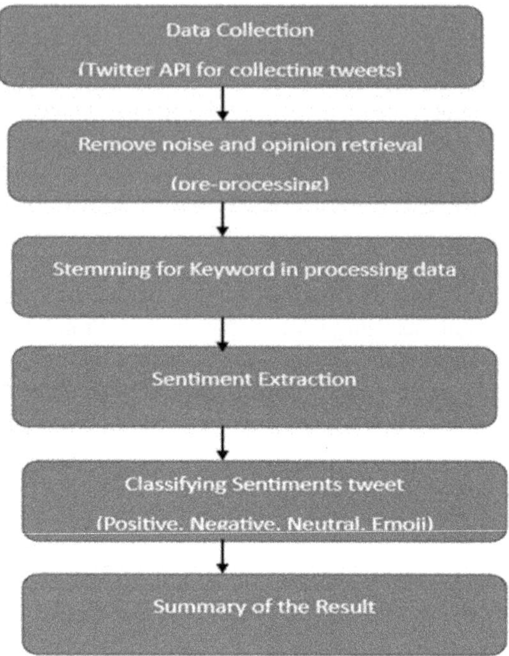

FIGURE 9.2 Flow process for proposed system.

system using linear regression and multinomial Naive Bayes to categorize the text in as positive or negative.

9.5.5 DATASET PRE-PROCESSING

To preprocess our dataset by removing unwanted data and noise. In the next step, we will do stemming and lemmatization of the dataset. Stemming is a progression of extracting or eradicating the last characters of a word, which frequently results in inaccurate interpretations and spellings. Considering the context, the lemma converts the word into its expressive base form, which is recognized as the lemma. Here, we use Porter Stemmer and Word Net lemmatizer. We will clean our dataset by removing abbreviations like @, #, *.%, $, &, etc. We will also use TF-IDF to find the furthest communal word used in the article or text. TF and IDF combine the two distinct metrics of TF and IDF. When there are numerous documents, TF and IDF are utilized. This is constructed on the concept that uncommon words reveal more about a document's content than terms that are frequently utilized throughout all publications. TF and IDF are calculated with the following formulas: where d represents a document, 'N' signifies the total quantity of credentials, and df is the quantity of credentials comprising the phrase t. TF-IDF is a word frequency statistic highlighting the most intriguing words. The scores indicate distinct terms in a given document in Figure 9.4.

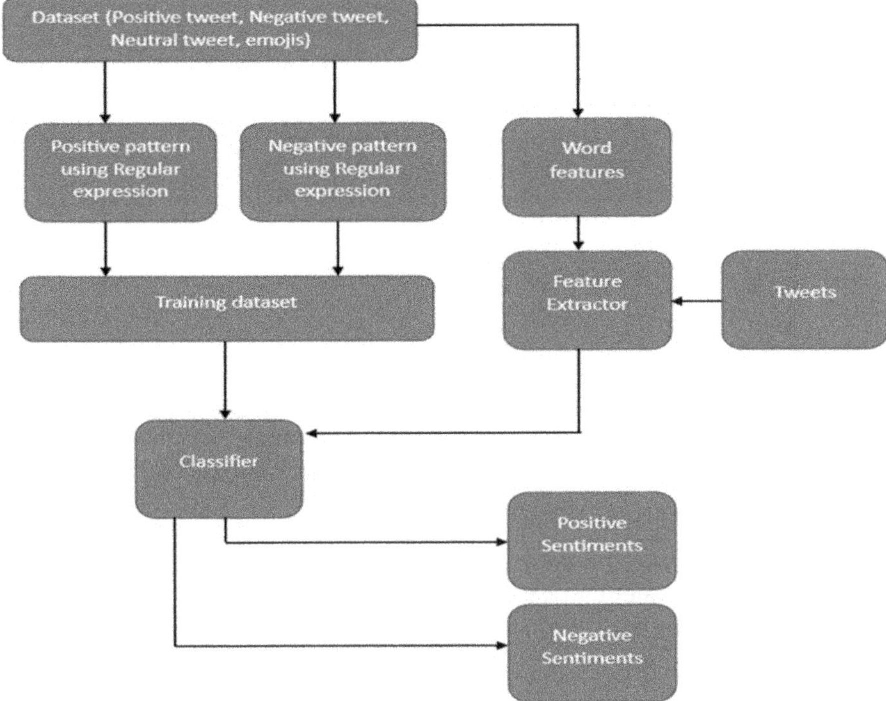

FIGURE 9.3 Proposed architecture.

FIGURE 9.4 Stemming of words.

We will also use TF-IDF to determine the most communal word incorporated in the file or text. TF and IDF combine the two distinct metrics to process further stages. When there are numerous documents, TF-IDF are utilized. This is constructed on the concept that uncommon words reveal more about a document's content than words that are frequently used across all publications. TF-IDF is calculated with the following formulas: where d signifies a document, "N" is the overall quantity of credentials, and df denotes the quantity of documents comprehending the tenure t indicated in Equations 9.1, 9.2, and 9.3. TF-IDF is a word frequency score that aims to highlight more fascinating terms. The scores have the effect of highlighting specific words in a given manuscript.

$$TF(t,d) = \frac{number\ of\ times\ t\ appears\ in\ d}{total\ number\ of\ terms\ in\ d} \tag{9.1}$$

$$IDF(t) = log\frac{N}{1+df} \tag{9.2}$$

$$\text{TF\_IDF}\ (t, d) = \text{TF}\ (t', d') * \text{IDF}\ (t') \tag{9.3}$$

9.5.6 Classification

The final step in the process is applying the classifier used here, which is Multinomial Naïve Bayes, and linear regression with regular expression. Multinomial Naive Bayes is a popular method for categorizing documents according to a statistical assessment of their material, as seen in Equation 9.4. It offers a feasible alternative to "heavy" AI-based semantic evaluation while considerably simplifying textual data classification.

The Bayesian probability $p1(C_k \mid W)$:

$$p1(C_k \mid W) = \frac{p(C_k) * p(C_k \mid W)}{p(W)} \tag{9.4}$$

The fundamental tenet of the Naive Bayes is that each characteristic in W separately influences the probability that S fits to C_k.

$p1(C_k \mid W)$ Equation 9.5 can also be given as:

$$posterior = \frac{prior * likelihood}{evidence} \tag{9.5}$$

The likelihood $(W \mid C_k)$ and estimated prior (C_k) both proportionally contribute to the outcome that C_k is the class of S.

Text search strings can be expressed using a language known as regular expressions (REs). RE allows us to compare or identify other strings or groupings of strings by encoding specified syntax into a pattern. Similarly, regular expressions are utilized to search texts in MS WORD and UNIX, as shown in Figure 9.5.

9.5.7 Result Analysis

Using regular expressions, we will find the common 100 words. Wh767hg5y6yt5545565e have written patterns that include positive emojis, positive words, etc.; here, we have created a list of complex patterns for positive text. Here, we are making an ML model using MultinomialNB(). Then, we have done TF-IDF vectorization of the text. We have taken the max feature to ten because we want to train our model on the ten

```python
# List of sample keywords
keywords = ["flight", "cancel", "time", "help", "hold", "plane", "call",
            "gate", "hour", "thank", "still" "😊", "😀", "👍", "❤", "😁",
            "😂", "😢", "👎", "😒", "😎"]
pattern = r'\b(?:' + '|'.join(keywords) + r')\b'
regex = re.compile(pattern, re.IGNORECASE)
# Example tweets
tweets = [
    "My flight was cancelled and I need help.",
    "I've been on hold for over an hour, can someone please help?",
    "The plane is still at the gate, not sure what the delay is.",
    "Thank you for the update on the flight time.",
    "I tried to call customer service but no one answered.",
]
for tweet in tweets:
    matches = regex.findall(tweet)
```

FIGURE 9.5 Cross model 1.

```python
import re

# List of positive emojis and words
positive_emojis = [
    '😊', '😀', '😁', '😄', '😃', '😉', '😇', '😍', '👍', '😎', '❤', '💛', '💚', '✨', '💯'
]

positive_words = [
    r'\bthanks\b', r'\bthank\s+you\b', r'\bawesome\b', r'\bgood\b', r'\bgreat\b',
    r'\bexcellent\b', r'\bamazing\b', r'\bhappy\b', r'\bfantastic\b', r'\blove\b', r'\bjoy\b'
]

positive_pattern = '|'.join(positive_emojis + positive_words)
positive_regex = re.compile(positive_pattern)
text = "I am so happy! 😊 This is fantastic! Thanks a lot! 👍"
matches = positive_regex.findall(text)
print("Positive matches found:", matches)
```

FIGURE 9.6 Cross model 2.

most occurring words. After training our dataset on MultinomialNB(), we print the accuracy of the model in Figure 9.6.

9.5.8 ANALYSIS

The results of our model outperformed previous ones significantly. A comparison between our developed models and multinomialNB is illustrated in Figures 9.7 and 9.8. While the current state-of-the-art precision value for multinomialNB stands at 0.82, our model demonstrates a precision value of 0.86. Furthermore, our model exhibits a recall value of 0.52 for positive and 0.90 for negative sentiment, representing a significant improvement compared to the state-of-the-art value of 0.22 for multinomialNB (as shown in Figures 9.9–9.12). Moreover, our model achieves an F1 score of 0.56, which is notably higher than the state-of-the-art score of 0.35 for multinomialNB shown in Tables 9.2–9.4.

9.5.9 RESULT BEFORE OUR APPROACH

Linear Regression

```
data = {
    'precision': [0.81, 0.82, 0.83, 0.82, 0.82],
    'recall': [0.99, 0.22, 0.60, 0.82, 0.90],
    'f1-score': [0.89, 0.35, 0.82, 0.62, 0.80],
    'support': [897, 258, 1155, 1155, 1155]
}
labels = ['Negative', 'Positive', 'accuracy', 'macro avg', 'weighted avg']
```

FIGURE 9.7   Positive and negative sentiments with various parameter for linear regression.

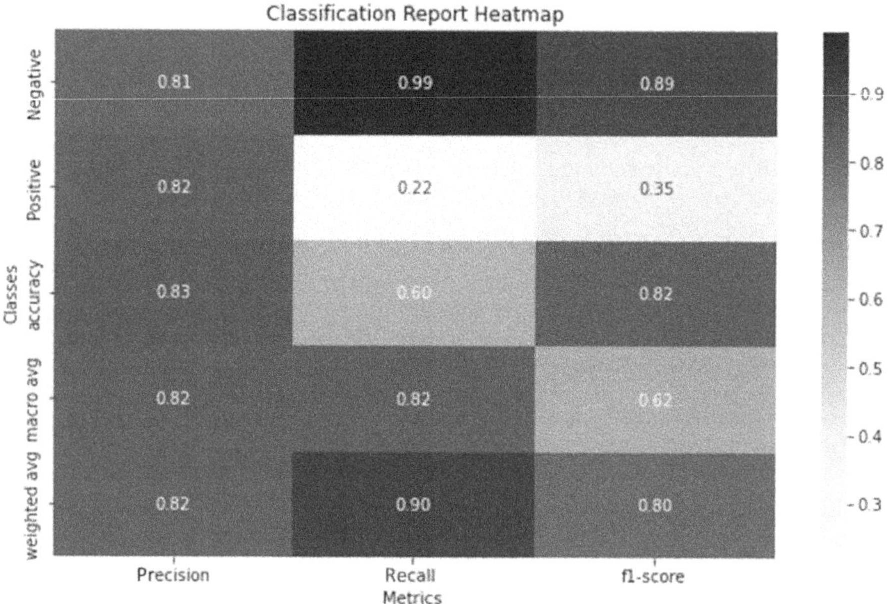

FIGURE 9.8   Positive and negative sentiments with accuracy for linear regression.

```
data = {
    'precision': [0.90, 0.60, 0.86, 0.75, 0.85],
    'recall': [0.93, 0.52, 0.86, 0.72, 0.86],
    'f1-score': [0.91, 0.56, 0.86, 0.74, 0.85],
    'support': [897, 258, 1155, 1155, 1155]
}
labels = ['Negative', 'Positive', 'accuracy', 'macro avg', 'weighted avg']
```

FIGURE 9.9   Positive and negative sentiments with various parameter for MultinomialNB.

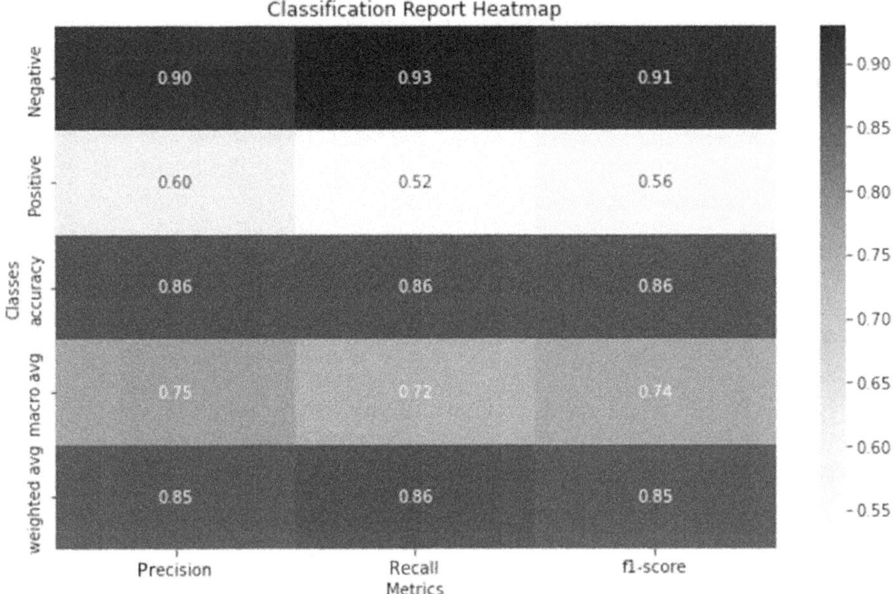

FIGURE 9.10 Positive and negative sentiments with accuracy for MultinomialNB.

9.5.10 Result After Implementing Proposed Approach

```
                precision    recall  f1-score   support

    negative       0.91      0.91      0.91       952
    positive       0.59      0.60      0.59       203

    accuracy                           0.86      1155
   macro avg       0.75      0.75      0.75      1155
weighted avg       0.86      0.86      0.86      1155

Confusion Matrix: [[867  85]
 [ 82 121]]
```

FIGURE 9.11 Positive and negative sentiments with various parameter for linear regression.

$$Accuracy = \frac{TN + TP}{TN + FP + TP + FN} \tag{9.6}$$

$$Precision(P) = \frac{TP}{TP + FP} \tag{9.7}$$

$$Recall(R) = \frac{TP}{TP + FN} \tag{9.8}$$

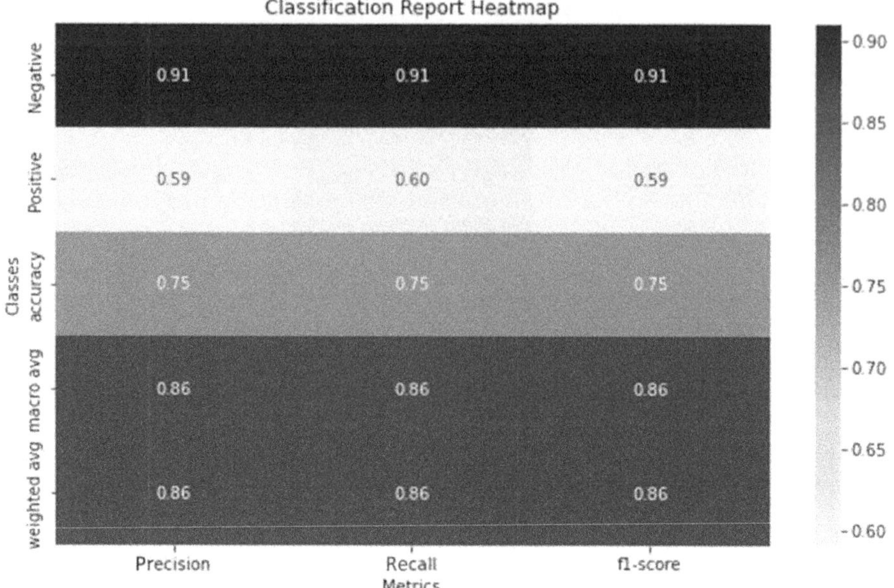

FIGURE 9.12   Positive and negative sentiments with accuracy for linear regression.

TABLE 9.2
Twitter Airlines Review

Dataset	Algorithm Used	Accuracy	
		Initial Accuracy (%)	Final Accuracy (%)
Twitter Airlines review	Linear regression	82.0	86.3
Twitter Airlines review	MultinomialNB	82.3	86.9

TABLE 9.3
For Positive Sentiment

	Precision	Recall	F1_score	Accuracy
Linear regression	59	60	59	86
MultinomialNB	60	52	86	86.9

TABLE 9.4
For Negative Sentiment

	Precision	Recall	F1_score	Accuracy
Linear regression	90	93	91	86
MultinomialNB	91	91	91	86.9

$$F1-score = \frac{Precision\left(P\right) * Recall\left(R\right)}{Precision\left(P\right) + Recall\left(R\right)} \tag{9.9}$$

Accuracy
The distinctiveness lies in the ratio of the combination of true negative (TN) with true positive (TP) divided by the sum of true positive (TP), false negative (FN), false positive (FP), and true negative (TN). The accuracy of the projected classification is achieved and detailed in Tables 9.2–9.4 and Equation 9.6.

Precision
It is distinct as the proportion of the quantity of tweets related to the quantity of inappropriate data retrieved. The performance of the projected classification accomplishes high when the rate of precision is high, and it is low when the precision rate is low, as depicted in Tables 9.2–9.4 and Equation 9.7.

Recall
It is distinct as the proportion of (TP)True positive by the sum of true negative (TN) and false negative (FN). The recall of the projected classification is accomplished and mentioned in Tables 9.2–9.4 and Equation 9.8.

F1 Score
It is distinct as the ratio of meticulousness multiplied by recall(R) and the sum of precision with recall. The Fl score of the projected classification is accomplished and mentioned in Tables 9.2–9.4 and Equation 9.9.

9.6 CONCLUSION

In this chapter, sentiment scrutiny of airline tweets consuming machine learning systems and regular expression has been training itself. Our model examines a total of 11541 tweets from 6 distinct airlines and we implement a technique of multinomial NB with regular expression. Also, the proposed concept in the chapter, which has much better accuracy as a result, is more efficient and has improved accuracy, which was successful. The retrieved characteristics are trained. And classified Twitter airlines review, got output as for initial accuracy for linear regression is 82% and final accuracy is 86.3%, for MultinomialNB, the output as for initial accuracy for linear regression is 82.3% and final accuracy is 86.9%. The result analysis of positive sentiment using linear regression, we achieved 59% for precision, 60% for recall, 59% for F1 score, and a final accuracy is 86%, and similarly in the result analysis of positive sentiment using MultinomialNB we achieved 60% for precision, 52% for recall, 86% for F1 score and accuracy is 86.9%. For the result analysis of negative sentiment using linear regression, we achieved 90% for precision, 93% for recall, 91% for F1 score, and a final accuracy is 86%, and similarly, result analysis of positive sentiment using MultinomialNB we achieved 91% for precision, 91% for recall, 91% for F1 score, and accuracy is 86.9%.

REFERENCES

[1] James Schaffer and Brandon Huynh, "An Analysis of student Behavior in Two Massive open online courses," *IEEE/ACM International Conference on Advances in Social Network analysis and mining (ASONAM)*, 2016.

[2] Kevin Andrey Ferreira and Liedy's del Carmen, "Analysis of Behavior of Customers in the Social Networks using Data Mining Techniques," *IEEE/ACM International Conference on Advances in Social Network analysis and mining (ASONAM)*, 2016.

[3] C. Verma, V. Stoffová, Z. Illés, S. Tanwar and N. Kumar, "Machine Learning-Based Student's Native Place Identification for Real Time," *IEEE Access*, vol. 8, pp. 130840–130854, 2020, DOI: 10.1109/ACCESS.2020.3008830.

[4] M. J. Adarsh and P. Ravikumar, "An Effective Method of Predicting the Polarity of Airline Tweets Using Sentimental Analysis," *2018 4th International Conference on Electrical Energy Systems (ICEES)*, 2018, pp. 676–679, DOI: 10.1109/ICEES.2018.8443195.

[5] F. Es-Sabery, A. Hair, J. Qadir, B. Sainz-DeAbajo, B. García-Zapirain and I. D. L. Torre-Díez, "Sentence-Level Classification Using Parallel Fuzzy Deep Learning Classifier," *IEEE Access*, vol. 9, pp. 17943–17985, 2021, DOI: 10.1109/ACCESS.2021.3053917.

[6] A. Ishaq, S. Asghar and S. A. Gillani, "Aspect-Based Sentiment Analysis Using a Hybridized Approach Based on CNN and GA," *IEEE Access*, vol. 8, pp. 135499–135512, 2020, DOI: 10.1109/ACCESS.2020.3011802.

[7] X. Lei, X. Qian and G. Zhao, "Rating Prediction Based on Social Sentiment From Textual Reviews," *IEEE Transactions on Multimedia*, vol. 18, no. 9, pp. 1910–1921, Sept. 2016, DOI: 10.1109/TMM.2016.2575738.

[8] Z. Chen, F. Lu, X. Yuan and F. Zhong, "TCMHG: Topic-Based Cross-Modal Hypergraph Learning for Online Service Recommendations," *IEEE Access*, vol. 6, pp. 24856–24865, 2018, DOI: 10.1109/ACCESS.2017.2782668.

[9] HT. Manshu and W. Bing, "Adding Prior Knowledge in Hierarchical Attention Neural Network for Cross Domain Sentiment Classification," *IEEE Access*, vol. 7, pp. 32578–32588, 2019, DOI: 10.1109/ACCESS.2019.2901929.

[10] K. Abdalgader and A. A. Shibli, "Experimental Results on Customer Reviews Using Lexicon-Based Word Polarity Identification Method," *IEEE Access*, vol. 8, pp. 179955–179969, 2020, DOI: 10.1109/ACCESS.2020.3028260.

[11] H. S. Lee, "Exploring the Initial Impact of COVID-19 Sentiment on US Stock Market Using Big Data," *Sustainability*, vol. 12, no. 16, p. 6648, 2020, DOI: 10.3390/su12166648.

[12] G. Xu, Z. Yu, H. Yao, F. Li, Y. Meng and X. Wu, "Chinese Text Sentiment Analysis Based on Extended Sentiment Dictionary," *IEEE Access*, vol. 7, pp. 43749–43762, 2019, DOI: 10.1109/ACCESS.2019.2907772.

[13] M. Rathi, A. Malik, D. Varshney, R. Sharma and S. Mendiratta, "Sentiment Analysis of Tweets Using Machine Learning Approach," *2018 Eleventh International Conference on Contemporary Computing (IC3)*, pp. 1–3, 2018, DOI: 10.1109/IC3.2018.8530517.

[14] W. Tai, T. Zhong, Y. Mo and F. Zhou, "Learning Sentimental and Financial Signals With Normalizing Flows for Stock Movement Prediction," *IEEE Signal Processing Letters*, vol. 29, pp. 414–418, 2022, DOI: 10.1109/LSP.2021.3135793.

[15] K. Cheng, Y. Yue and Z. Song, "Sentiment Classification Based on Part-of-Speech and Self Attention Mechanism," *IEEE Access*, vol. 8, pp. 16387–16396, 2020, DOI: 10.1109/ACCESS.2020.2967103.

10 Smart Workspace Automation

Harnessing IoT and AI for Sustainable Urban Development and Improved Quality of Life

Sapna R., Preethi, Pavithra N., Manasa C.M., and Raghavendra M. Devadas

10.1 INTRODUCTION

Several individuals spend a significant amount of time in workplaces these times. Everybody should feel at ease in the office since the working environment seriously influences how successfully employees accomplish their tasks. So, convenience is essential and required in the workplace. When technology was at its greatest in the past, it meant having an electronic pager and a landline connection; today, it means having an electronic tablet linked to the web. A smart office is one that empowers life with ease and comfort for its people and clients, empowering them yet enhancing their capability to keep themselves connected to one another. Organizations are harnessing the power of modern technology and implementing effective strategies to drive staff efficiency and output. As the physical barriers are being breached, a multi-faceted and competitive atmosphere prioritizing innovation and creativity is forming. Intelligent growth zones are rapidly emerging worldwide, so smart offices quickly become necessary.

A "smart workplace" ensures IT resources and infrastructural facilities are used as effectively and efficiently as possible. In other terms, offices are automated in the current generation of information technology. An accessible, innovative technical setting is required. Thus, office automation replaces processes that are more accessible and encourages open information exchange, which creates the opportunity to have a substantial effect on how industries as well as enterprises operate. The system's usage of various communication tools and efficient automated systems demonstrates the favorable effects on a company's or lender's long-term growth. The benefit of a smart office is eliminating internal reporting procedures, such as employee arrival and departure times, through an open office layout. Improving cooperation and communication can raise efficiency, which influences the outcome.

DOI: 10.1201/9781032623276-10

A smart workplace needs to be developed to maximize employee and worker potential. This is not a supernatural event; it is just revolutionary innovation and creative thinking that better meets people's requirements. Direct conversation and simple recordkeeping are among the benefits of office automation. One of the key drivers of worldwide energy consumption is the building sector. It uses almost a third of all the energy consumed. Like a smart house, a smart building uses sophisticated sensor data collection and analysis to maximize comfort, efficiency, and safety. Complicated machine devices, sophisticated control systems, and other elements are found in modern buildings to increase occupant safety and efficiency.

A fantastic system of linked subsystems might be considered a smart building. All a building's systems and appliances must be connected to be considered a smart building. The managers can utilize it to view information and quickly and accurately establish judgments. Smart buildings can be made using a variety of methods. The automated centralized control of an office infrastructure's various components, such as light, heating, ventilator, cooling systems, and other systems, is accomplished via an automatic building management system. The objectives of smart building systems are comfortable structural framing functioning, lower energy consumption and operating costs, and extended utility developmental stages.

By controlling the power consumption in business buildings, considerable energy can be saved. These problems lead to the concept of the "smart office." This smart office design concept can be used for the entire structure to use less energy. In these days, security is essential. Due to the introduction of rapidly developing technology, users now need strong security solutions. The systems need to be very secure. These identifying techniques include user password-based systems, ATMs, other smart cards, and many others. Nevertheless, hacking assaults, theft, and credential forgetting render these systems insecure.

Radio-frequency identification (RFID) has also been employed for security reasons. However, RFID has the drawback that it can be hijacked, or another individual can enter using that ID. This is definitely not safe. Despite all these flaws and errors, the most effective and dependable security solution is a biometric or biometric authentication-based identity.

That would save energy and improve employee satisfaction. A complete smart office system with targeted lighting, heating, monitoring, and alerts is constructed using this approach. Several strategies are used in smart energy solutions to create energy savings that benefit from organizational conduct. This strategy, which has been implemented, exemplifies how to cut back on energy use by realizing the significant influence that people have on it.

Due to the extensive use of innovative technologies in professional settings, smart office concepts have gained popularity. Even though companies have been implementing smart office ideas to give their customers practical and efficient workspaces [1], the studies have given less attention to the user's perspective. It is well known that relevant issues and aspirations should be considered when developing workplace settings [2]. Still, it is unknown what intentions and interests people have for automation and control ideas, as well as what specific features justify smart office concepts and differentiating themselves from other non-smart workplace types. To comprehend the user requirements for smart office spaces, empirical information would be of great social and scientific relevance.

10.2 STATE OF THE ART

This section examines how the Internet of Things might boost production, worker productivity, and safety while enhancing performance evaluation procedures. It looks at how Internet of Things (IoT) can deepen the relationship between management and employees. System of work at manufacturing facilities. After reading the relevant literature, IoT aids in observing and controlling worker behavior, enhancing performance, and granting workers more independence. Analyzing IoT's role in worker safety and productivity is the first step in understanding how it may benefit workers. Organizations may not even be aware of all the safety challenges they face. IoT aids with the use of precise sensors. These problems are identified, advance warnings are given so that action can be taken, and rescue efforts are facilitated. Tracking workers' vitals and making suggestions in line with them not only helps to improve safety standards but also worker productivity. Building this dynamic workspace would be practically impossible without the IoT, so we explore how it works. We also examine alternative IoT-based performance evaluation methods. Conventional techniques of measuring employee performance are occasionally prone to manager biases and are only undertaken occasionally. IoT, on the other hand, provides constant feedback and is impartial. Conventional HR practices will not be fully abandoned; rather, they will be strengthened with the help of consistent, dependable feedback.

We can answer the crucial queries of how IoT can assist and how it is used in human practices by going over these two subjects. This also establishes the framework, for instance, in which IoT is employed to track employees' geolocations when required to precisely understand how much time each worker contributes to work. Reinforce them with the aid of ongoing, accurate feedback. The 4th Industrial Revolution is the present phase of technological progress characterized by an emphasis on AI-driven technologies. Data-driven AI technologies are currently fostering an environment of Industry 4.0, which follows the 3rd Industrial Revolution launched by the Internet and mobile Internet. Industry 4.0 describes the current technology trend to automate processes and interchange information. Numerous inexpensive tools can gather real-time data and send it through the IoT. As a result, using a remote control is possible.

The IoT makes it possible for machines and devices to communicate with one another and generate huge quantities of data, potentially revealing significant information in a few service sectors. We can create integrated semantic systems that enable semantic interoperability by fusing the IoT context with semantic technologies. The development of research systems combines embedded IoT, which is used to remotely operate home appliances, particularly electronic equipment, with embedded ESP8266, which acts as a control with real-time firebase. ATMEGA16 microcontroller and 20

18 Fourth International Conference on Computing, Engineering, and Design [3]. The local and global Message Queuing Telemetry Transport (MQTT) servers can be used to construct a remote temperature monitoring system that employs the MQTT protocol and a database that serves as a mobile Backend Service to store data [4]. MQTT server is used, anyone with an internet connection can access the temperature from anywhere at any time. The IoT technology can be utilized to protect homes from unforeseen issues, including fire, theft, temperature conditions, and motion gestures [5].

One of the pioneering studies to emphasize the significance of the sustainable development of IoT and how innovative technology should not be a goal in and of itself is that of the authors in [6]. Hoang et al. [7] have evaluated how various Information and Communication Technology (ICT) solutions could be leveraged to lower energy usage. The studies by Adhikari et al. [8] have more recently discussed incorporating sustainable energy sources into smart developments and any potential difficulties. As an expanding dissertation topic on the need aspect of a smart grid, the hurdles to the deployment of electric vehicles in relation to energy demand have been exhaustively analyzed by the researchers in Sanguesa et al. [9]. Analogously, the examinations conducted by Omitaomu et al. [10] assessed the matter, albeit in the specific domain of smart cities. The studies by Yigitcanlar et al. [11] have offered a thorough assessment on the subject, and numerous AI applications have been created for the smart grid. Still, it has not yet been determined if the adoption of AI can address environmental issues. The researchers in Crossley and Beviz [12] analyzed the general limitations of AI in smart buildings.

10.3 USING AUTOMATION TO ADDRESS ENERGY ISSUES

Electricity production, public transit, and construction will all be considered, as they are the main factors that determine emissions of greenhouse gases. Figure 10.1 illustrates the key technology components that deal with energy in Smart Office.

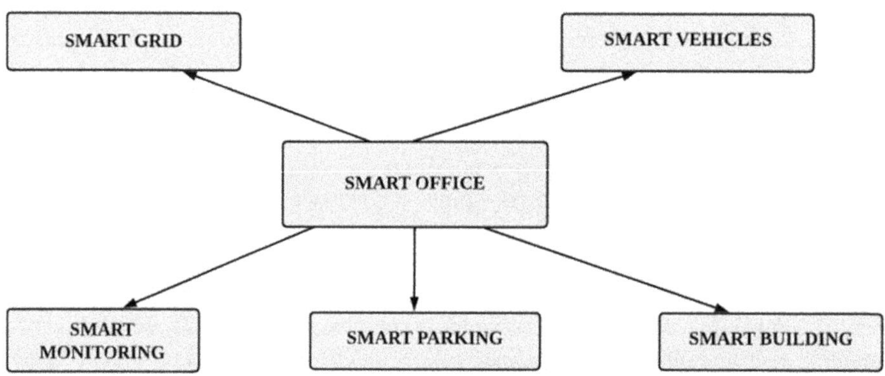

FIGURE 10.1 Smart office.

10.3.1 SMART GRID

A smart grid, a power arrangement that relies upon sensors, communications, controllers, and information systems, uses IoT ideas to harmonize the power grid's current activities and make them more efficient and dynamic in their administration. The IoT, specifically wireless sensor networks, tracks solar energy production. In the agricultural sector, IoT technology has advanced to facilitate the identification of viable agricultural land and the selection of suitable crops by utilizing Raspberry Pi. Within the healthcare domain, IoT is employed to monitor cardiac activity precisely, specifically focusing on heart rate.

The idea of a smart grid is inextricably linked to the emergence of rapidly changing sustainable assets with transferring belief toward alternatives to fossil fuels, all while keeping supply and demand manageable [13]. A bi-directional power flow is a common cornerstone for recognition in a smart grid, which implies that customers are also energy manufacturers, abandoning the standard function of inert users [14]. The main challenge for the smart grid is the networks themselves, as the majority were built to take in massive fossil fuel-powered facilities, supplying power to customers only in one way [15]. Figure 10.2 shows a general view of a typical power grid. Three vital elements required for this bidirectional flow are demand-side management, storage technology, as well as real-time infrastructure management. All these elements promote technical assistance. Four types, such as load prediction, power grid stability evaluation, defect analysis, and security concerns, have seen the most use of AI [11].

As evidence for the latter group, the incorporation of information technologies into the electric grid itself could create security issues because they can facilitate hacks meant to interfere with regular, daily operations. Artificial intelligence (AI) techniques were developed to recognize when such assaults occur. The development of a smart grid can be seen in the context of a smart workplace as a basic building block for addressing the environmental concerns and resource scarcity that are becoming increasingly widespread globally.

Variable pricing is one method used to match energy supply to demand. For example, a rise or drop is synchronized due to a rise or fall in requests made for the energy. As a result, during peak times, the cost is advanced even though energy plants are operating at near-maximum ability, whereas during normal hours, the cost is less. This means clients are urged to utilize equipment when energy demand is low. It's true that demand response efforts are designed to keep customers by providing a monetary reward in the form of lower electricity bills. These efforts' versatility is the capable of accommodating the variations of renewable energies and enabling their

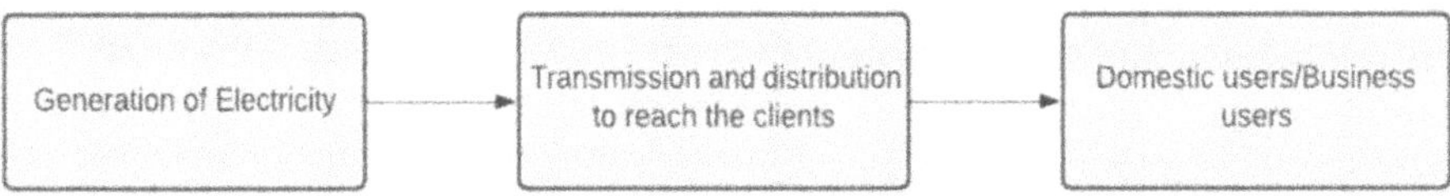

FIGURE 10.2 A general view of a typical power grid.

greater depth [16]. AI-powered solutions have been created to help such practices in this scheme. Boza and Evgeniou [17] suggested an RL algorithm to stabilize service suppliers' profits and clients' savings in order to attain power system reliability. Data centers are a potential load management sector where AI can be used [18]. The enormous volume of data a smart office should produce is among its most crucial features [19]. Nevertheless, the establishment and growth of such a smart workplace may face environmental challenges related to the collection and storage of data. Data centers are energy-intensive businesses that consumed 1% or more of the world's electricity in 2018. Their efficiency is based on the consolidation of multiple servers. According to the research by Crawford [20], data centers will use 2.13% of the total electrical supply by 2030 in 2021. Mytton [21] reports a warning against statements made by businesses that they are carbon neutral because this may be the result of buying carbon credits instead of using renewable sources of energy.

From an environmental standpoint, the sustainable development of data centers is inextricably linked to the use of renewable resources to influence their operations. It is crucial to note that exorbitant water usage is a related problem because water is used to keep down data centers, which can have a significant role in energy utilization. Yet, it is also employed in oblique ways to generate power [22]. This is significant to the water shortages due to climatic change, as it has been disclosed that the global water supply, such as snow and ice, has reduced by 1 cm per annum over the last 20 years [23]. Although service providers and lawmakers have taken steps to reduce these environmental issues [22], optimizing energy usage can help to solve this challenging problem. Li et al. [23] recommended an algorithm to optimize the data center. Cooling systems and reducing associated costs. A related real-world AI application in data centers is the ML method proposed by DeepMind for Google, which reduced the expanse of energy needed for the cooling system by 40%. According to reports, Google and DeepMind will work together to make the technology accessible to commercial organizations [24,25].

Energy storage is another method for achieving both supply and demand stability. There are numerous types of energy storage systems, including mechanical (like flywheels), electrochemical, thermal, and chemical. They may also be split into two groups based on where they apply. Grid-scale implementations appropriate for mass storage are accessible. Demand-side applications, on the other hand, are confined to small regions. Demand-side storage is frequently used with on-site energy production, as is typically with some sustainable energy sources, like wind and solar. A wide range of storage technology applications for AI can be thought of. AI has been used, for example, to create and evaluate the storage systems themselves, such as when choosing the requirements for lithium-ion batteries to increase their individualism. An algorithm was created to evaluate the suitable location and measurements of an energy storage system in the context of improving management in a business significantly an electric grid using renewable energy sources, as doing so should improve reliability and minimize power shortfalls [26].

10.3.2 Smart Monitoring of Employees

The stimulation of various electronic monitoring structures today makes it possible to improve management in a business significantly. Various firms are increasingly adopting technology to observe employees and improve staff effectiveness electronically. This is correlated to the enhancement of any business. This study aims to examine how several employee performance indicators are affected by electronic surveillance. Three different organizations' present surveillance systems have been noticed, and for the research, 20 staff members from each company were chosen as a sample. It has been found that these systems heavily rely on the employees' logging in and out times and that their punctuality and sincerity significantly affect both their individual and collective performance. To increase the productivity rate of the company and employees, no surveillance monitoring has ever been established that can track the complete activities of the staff members on the resources in office time. To effectively monitor any computer system during working hours, the solution provided by Chinyere and Chiemela [27] analyzes the required observational parameters and incorporates a keylogger into the system.

There are two distinct types of keyloggers: hardware-based and software-based keyloggers. Software keyloggers can be installed on a computer either locally or remotely. While hardware-based physical keyloggers are simple to find whether a user finds what devices/resources are linked to the computer, software keyloggers are invisible to the human eye. Software-based keyloggers can reproduce using a virtual system, hypervisor-based or virtual system manager, operate as a keyboard driver, or use other methods to access the operating system of the targeted machine [27].

Bernstrøm and Svare [28] examined the covert measures employers take to monitor their workers' behavior. E-monitoring and surveillance are the main approaches employed. Although it has been noted that the system in use is paradoxical, employers' goals are to accomplish their specific objectives and preserve the system's performance, effectiveness, and data protection. Practically, it becomes a disruption to their employees' privacy. According to research, more organizations are implementing this type of surveillance since the effects of this realization lead to expansion. The author further concluded that employers should try to uphold appropriate corporate conduct to ensure productive management.

Jeske and Kapasi [29] examine workplace surveillance efficacy and strive to distinguish between command and monitoring, as well as between fundamental inspiration and proficiency. Additionally, structural equation modeling was used in this study to examine the anticipated Norwegian employee sample samples. The method employers use to introduce and administer monitoring to their workforce will determine if it has a positive or negative effect on them. According to the findings, workers feel less convicted when they know they are being watched. However, the study reported in the chapter shows that employees are still less motivated and have less faith in the system, even once a surveillance system is in place. A system is necessary for employees to better understand how security measures are implemented.

Ahmad et al. [30] examined the information found in the electronic performance monitoring method. The primary focus of this study is on how data are generally used following the use of monitoring techniques. Surveillance to increase security and monitor employee performance may help the organization expand. Still, it has also been shown to occasionally breach employees' privacy and hurt how they view the company. This indicates that the distinction between the methods employed to assess the outcome is an important consideration for management edification and the formulation of moral judgment.

Vujović and Maksimović [31] suggested authoring a report to elaborate on the influence of performance monitoring, information/data security, and additional social immersion coaching elements on employees' behavior regarding safety assertion. The safety pledge represents the actions taken to protect employees' diligent work and information arrangements. The behavior is very desired because it relates to a personal aspect within the context of information safety. The authors concluded that implementing information security monitoring can enhance the cultural behavior of safety assertion. The study concluded that when employees' behavior is deemed inconvenient, safety procedures must be abandoned.

Many earlier studies on monitoring focused on the benefits and drawbacks of e-monitoring implementation and employees' reactions. Many businesses have introduced e-surveillance without the knowledge of their employees, which has a negative effect and makes them worry about their privacy. Most systems use monitoring systems to keep track of employees' time records. By utilizing the biometric technology built within their system, practically every firm heavily relies on this strategy. As a result, a system that can monitor employees' computer use while they are at work is needed, in addition to being able to record their arrival and departure times.

10.3.3 Smart Office Building with Controlling Lights, Fans, and Other Electronic

A Building automation system lowers energy usage and makes utilizing various appliances more convenient. Thanks to the energy-saving strategy, building automation today makes a living quite simple. All electrical and technological home appliances and devices will be controlled remotely through a wireless connection. Individual control device systems, distributed control device systems, and centrally controlled device systems are all possible configurations for automation systems [32]. The simplest home automation devices are individual control devices, which are programmable gadgets that may be configured to user preferences. On the other hand, distributed and centrally controlled systems, which build automation systems, have remote communication and device control capabilities. The distinction between distributed and central systems is that the latter has a controller to centrally control all the devices, while the former does not. The disadvantage of a centralized control system is the breakdown when the controller breaks down [32].

Commercial building automation systems are pricey because they require specialized hardware, components, and installation. Utilizing open-source platforms and IoT sensors to perform specialized automated home systems that is tailored to the demands of the end user is an alternative strategy. The cost of IP-based devices and customer inexperience with open-source home automation systems are the two main drawbacks of this strategy. This automation needs to install multiple such pricey IP-based devices, which are out of the price range of low-income homes.

The image shows that the system's primary implementation components are sensors, regulating mechanisms, and actuators. The sensors' data regarding light, motion, temperature, and other sensory features is transferred to the primary controlling devices. Various sensors, such as photodetectors, level sensors, pressure transducers, transformers, infrared sensors, and temperature sensors or thermistors, require additional signal processing apparatus to communicate with the main controller.

Controllers are designed to be coupled to control devices, such as programmable logic controllers, which get data from sensing devices and perform operations upon them in accordance with a program. This software could be altered by the load operations. Different analog or digital inputs and output units, such as sensing devices, actuating devices, and other devices can be connected to the programmable controller.

The last devices to regulate the appliances are called actuators, which comprise limit switches, relays, motors, and other controlling mechanisms. In the home automation system, connectivity is essential for remote access to these operations. Additionally, this smart home system provides scheduling, power features, and constant surveillance via video surveillance with cameras. This is the best treatment— even for the old. A simple office building automation system is shown in Figure 10.3.

Typically, it takes place in an automated home setting. It can be divided into several sorts. A smartphone can be used to control some of them remotely. Second, it can command a motor or actuator that assures security, like a door lock. Nowadays, many building automation systems are mostly Android-based, allowing a single smartphone to manage all household appliances. Arduino, relay modules, Wi-Fi (IEE802.11 b/g/n)/4G, 5G, or Bluetooth modules can be used with a PC or microcontroller to quickly create a home automation prototype.

Relay channel count is determined to quickly create a home automation prototype based on how many appliances you want to connect to the automation system. An Arduino board can accept up to 12 V of power. It might also be powered by a 9-V DC battery. Arduino's 5 V/3.3 V output pin can power Bluetooth and the relay module.

The most common asynchronous transfer method utilized by the Bluetooth module is the universal asynchronous receiver/transmitter (UART) mode with a band rate of 9,600–115,200. Use this mode to communicate with another device, such as a laptop or smartphone. In this instance, a native Android app was utilized to link the Bluetooth module with the Android phone and control the appliances.

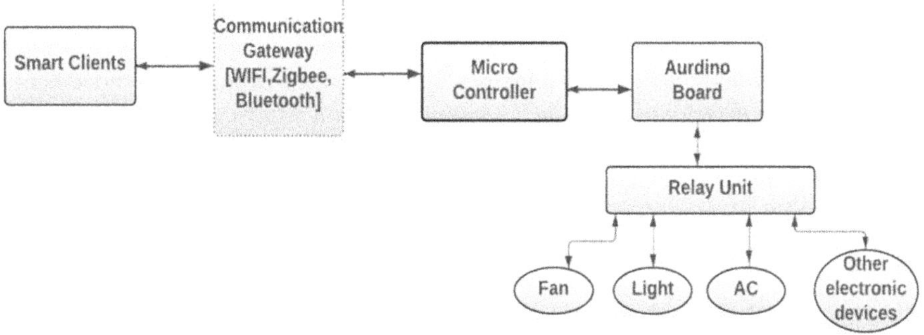

FIGURE 10.3   An office building automation system.

The relay primarily manages electrical devices, including fans, lights, refrigerators, and air conditioners. It offers total isolation between the high-voltage line and Arduino or another module. Its primary coil is agitated by delivering 5 V to the coil, which is mostly applied from the Arduino board.

10.3.4　Smart Vehicles with Automated Parking

The automated parking system combines software, a mobile application, and transportation instruments for the best possible user experience. This system is made up of a range of electromechanical devices and robots that are controlled by computers. Without the aid of a person, vehicles are stored, retrieved from, and placed into open multi-depth parking spaces using conveyance equipment, which can move both horizontally as well as vertically. The system under consideration is characterized by its electromechanical transport apparatus, designed within a closed structure that may consist of a single level or multiple levels. It facilitates movements across multiple dimensions, directions, and depths, indicating a versatile and dynamic range of transport capabilities. The system incorporates designated rooms for entry and exit, emphasizing its functional and organized design to facilitate seamless transitions and accessibility. The combination of electromechanical transport, closed structure, and multi-dimensional movements suggests a sophisticated system tailored for efficient and controlled transportation within various environments.

The automated parking garage has no humans driving or walking through it. High-density parking and effective land use are made possible by this. The system effectively uses the building by not requiring ramps, turning radii, or pedestrian walkways, freeing up valuable space for additional units or amenities.

The automatic parking system functions in contrast to a standard one. It is almost like a valetless valet parking service; it is speedier and requires less driver involvement. The automated parking system has several bay rooms for parking and riding away vehicles. The car's driver approaches the harbor and ceases at the arrowed sign. When a rolling door gives way, the driver can take the vehicle into the bay, where the sensing device measures the vehicle's measurements and then instructs the driver on how to position the car properly on a screen in front of them. When the car has reached the right position and is prepared for parking, the driver sees a notification on the screen.

After securing the car with the key and exiting the vehicle, the driver drives to finish the parking process. The parking operation is started with a smartphone app, a card swipe at the payment terminal, or a pay ticket. When the app verifies the parking request, the bay door closes, and sensors scan the area to verify that no movement is observed on the exterior of the vehicle. A shuttle system removes the vehicle from the bay and stores it in the right area based on dimensions. The Bay room is now ready for another vehicle to be parked or retrieved from safekeeping.

10.3.5　Industry Revolution

The continued Industrial changes and growth of Industry 4.0 from the start of the Industrial Revolution 1.0 (1760–1840) have caused a change in basic assumptions in healthcare centers provided globally. Industry 4.0 combined developments in computing techniques with the emergence of cyber systems, the IoT, cloud computing, cognitive computing systems, and artificial intelligence (AI) technical platforms. Since its introduction in 2011, innovations in Industry 4.0 have revolutionized caring for patients at every stage, including integrating medical facilities, accelerating the growth of groundbreaking treatments, and prompt detection.

Unfortunately, Industry 4.0 did not meet the increasing demand for "Personalized care" despite recent automation and high efficiency satisfying the requirement for production in large quantities with reduced human effort using intelligent production methods. Industry 5.0, a concept first suggested in 2015, has made it possible to personalize products and integrate more human intelligence into manufacturing. Enhancing the interplay between smart machine systems, human intelligence, and robotic technology is anticipated to design and deliver individualized care and responsibility for patients with orthopedic disorders. This helps in building on the sophistication of Industry 4.0 technologies.

Industry 5.0, unlike its predecessor Industry 4.0, shifts its focus from solely maximizing production efficiency through automation to prioritizing the human element in manufacturing. It emphasizes the development and implementation of cutting-edge technologies, such as the IoT and big data analytics, to enhance human capabilities and improve job satisfaction. While Industry 4.0 focused on replacing human labor with robots and intelligent systems, Industry 5.0 aims to create collaborative environments where humans and machines work together synergistically, leveraging each other's strengths. With proof of recent research and technological development, industry 5.0 will advance marketably. Over the last five years, the manufacturing sector has witnessed significant advancements and transformations driven by key technologies. Artificial intelligence (AI) and machine learning (ML) have played pivotal roles in optimizing production processes, enhancing predictive maintenance, and improving overall efficiency. Collaborative Robots (Cobots) have been integrated into manufacturing workflows, fostering human–robot collaboration to streamline tasks and increase productivity. The Internet of Everything (IoET) has interconnected devices and systems, creating a smart and interconnected manufacturing environment. Blockchain technology has been employed to enhance supply chain transparency, traceability, and security. Additionally, adopting digital twins has become prominent, enabling virtual replicas of physical assets to enhance monitoring, analysis, and decision-making in the manufacturing sector. Collectively, these technologies signify a paradigm shift toward more intelligent, connected, and data-driven manufacturing processes. Their use cases will change, and new implementation areas will emerge that are essential to defining Industry 5.0. AI has been integrated into every stage of the manufacturing value chain. AI-enhanced human machines are a crucial development that will reestablish AI's standing in Industry 5.0. The level of interaction between humans and robots has increased from simple to complex, encompassing the use of data to verify, manufacture, and operate autonomous vehicles and the development of sophisticated storage facility-picking robots capable of classifying commodities based on the order in which they are obtained

With the confluence of IoT, big data analytics, and AI, the Smart Factory was born. Industry 4.0 marked the beginning of the IT-OT combo, which is still expanding to an extent to become IoET. IoET, taken as a whole, refers to various intricately coupled devices.

In the next decades, these developing technologies are anticipated to grow in a number of important fields, including:

1. AI-ML: Increased Intelligent Automation will boost office and shop floor productivity. Businesses place a high priority on quality management, and AI will help them do that by facilitating decisions in less time. For instance, process manufacturer Koch collaborates with Independent Software Vendor (ISVs) headed by AI, such as C3.AI, to deploy AI throughout the organization. This will enable managing energy use, planning and scheduling manufacturing, and optimizing inventory.
2. Cobots: The employment of cobots in production and warehouse management will boost factory productivity, leading to improved uniformity and precision in product production. ABB, the first company to offer cobots, is leading technological breakthroughs with its 46 years of cobot-building experience. Cobots are being introduced to replace manual labor on the shop floor for Automated Guided Vehicles (AGV) mounting, assembly line automation, simple packaging, and industrial testing.
3. IoET: IoT is gradually evolving into IoET, and connected industrial equipment is assisting producers in reducing costs, increasing asset productivity, lowering downtimes, and establishing realistic supply chain visibility. A notable example is Koch, which is utilizing invisible, IoT-enabled tapes from US-based ISV to make its factories smarter. This clever wireless tape transmits data like location, position, and temperature while monitoring valuable cargo and equipment.
4. Blockchain: Assuring operational transparency and enhancing supply chain visibility, asset monitoring, and operational enhancements for manufacturers are all aspects of operations in which blockchain technology is gaining prominence. This development is not just limited to the Banking, Financial Services, and Insurance (BFSI) sectors.
   With Honeywell, this development has already taken a step forward. To replace manual processes and paper use, Honeywell Aerospace is assiduously investing in Blockchain to list products for sale, track stakeholder interactions, and keep track of transactions.

10.4 CONCLUSION

The existing smart workplace principles are expanded upon in this chapter from the perspective of users. Because cutting-edge technology and design might create an expectation of modern workplaces with smart office ideas, the requirement for ongoing improvement of the workplaces must be considered from a management perspective. This study recommends that a precise statement of the intelligent idea should be supplied to the consumers, referring to the specific qualities, as various users may have varied viewpoints and aspirations for smart ideas. With the smart workspace, employees can manage the equipment throughout the office to cut down on electricity waste. For the upcoming development, it is anticipated that a better system will be developed and created with the addition of more and different outcomes to allow for greater enhancement in attempting to control other equipment present throughout the workplace. It can also be created with a data warehouse within every individual's smart working space.

All new construction will soon include sensing and communication systems to give occupants the desired convenience. As a result, it will impact how buildings are designed in the future, possibly even turning non-residential facilities like hotels and airports into intelligent structures. There will be a multitude of cutting-edge life cycle models for smart buildings. To give users of the smart building a richer user experience, all units should be flexible enough to accommodate technological improvements.

REFERENCES

[1] Bodker, S. "Rethinking technology on the boundaries of life and work," *Personal and Ubiquitous Computing*, vol. 20, no. 4, 2016, pp. 533–544, DOI: 10.1007/s00779-016-0933-9.

[2] Kim, J., De Dear, R. "Workspace satisfaction: The privacy-communication trade-off in open-plan offices," *Journal of Environmental Psychology*, vol. 36, 2013, pp. 18–26.

[3] Alipio, M.I. Development of Smart Indoor Workplace System Using Decision Tree Algorithm. In *2021 IEEE International Conference on Internet of Things and Intelligence Systems (IoTaIS)*, 2021, pp. 196–202. IEEE.

[4] A. Jaribion, S. H. Khajavi, J. Holmström, "IoT-enabled workplaces: a case study of energy management and data analytics," *IECON 2019 – 45th Annual Conference of the IEEE Industrial Electronics Society*, 2019, pp. 5325–5330, DOI: 10.1109/IECON.2019.8927003.

[5] Ahvenniemi, H., Huovila, A., Pinto-Seppä, I., Airaksinen, M. "What are the differences between sustainable and smart cities?" *Cities*, 2017, 60, 234–245.

[6] Kramers, A., Höjer, M., Lövehagen, N., Wangel, J. "Smart sustainable cities – Exploring ICT solutions for reduced energy use in cities," *Environmental Modelling & Software*, 2014, 56, 52–62.

[7] Hoang, A.T., Pham, V.V., Nguyen, X.P. "Integrating renewable sources into energy system for smart city as a sagacious strategy towards clean and sustainable process," *Journal of Cleaner Production*, 2021, 305, 127161.

[8] Adhikari, M., Ghimire, L.P., Kim, Y., Aryal, P., Khadka, S.B. "Identification and analysis of barriers against electric vehicle use," *Sustainability*, 2020, 12, 4850.

[9] Sanguesa, J.A., Torres-Sanz, V., Garrido, P., Martinez, F.J., Marquez-Barja, J.M. "A review on electric vehicles: technologies and challenges," *Smart Cities*, 2021, 4, 372–404.

[10] Omitaomu, O.A., Niu, H. "Artificial intelligence techniques in smart grid: a survey," *Smart Cities*, 2021, 4, 548–568.

[11] Yigitcanlar, T., Mehmood, R., Corchado, J.M. "Green artificial intelligence: towards an efficient, sustainable and equitable technology for smart cities and futures," *Sustainability*, 2021, 13, 8952.

[12] Crossley, P., Beviz, A. "Smart energy systems: transitioning renewables onto the grid," *Renewable Energy Focus*, 2010, 11, 54–59.

[13] Lund, H., Andersen, A.N., Østergaard, P.A., Mathiesen, B.V., Connolly, D. "From electricity smart grids to smart energy systems—A market operation based approach and understanding," *Energy*, 2012, 42, 96–102.

[14] Blackmon, D. Renewables Won't Save Us If the Electric Grid Is Not Ready. *Forbes*. 2020.

[15] O'Connell, N., Pinson, P., Madsen, H., O'Malley, M. "Benefits and challenges of electrical demand response: A critical review," *Renewable and Sustainable Energy Reviews*, 2014, 39, 686–699.

[16] Lu, R., Hong, S.H., Zhang, X. "A dynamic pricing demand response algorithm for smart grid: reinforcement learning approach," *Applied Energy*, 2018, 220, 220–230.

[17] Boza, P., Evgeniou, T. "Artificial intelligence to support the integration of variable renewable energy sources to the power system," *Applied Energy*, 2021, 290, 116754.

[18] Hashem, I.A.T., Chang, V., Anuar, N.B., Adewole, K., Yaqoob, I., Gani, A., Ahmed, E., Chiroma, H. "The role of big data in smart city," *International Journal of Information Management*, 2016, 36, 748–758.

[19] Koot, M., Wijnhoven, F. "Usage impact on data center electricity needs: a system dynamic forecasting model," *Applied Energy*, 2021, 291, 116798.

[20] Crawford, K. *The Atlas of AI: Power, Politics, and the Planetary Costs of Artificial Intelligence*. New Haven, CT: Yale University Press, 2021.

[21] Mytton, D. "Data centre water consumption," *npj Clean Water*, 2021, 4, 11.

[22] Tiefenbeck, V., Wörner, A., Schöb, S., Fleisch, E., Staake, T. "Real-time feedback promotes energy conservation in the absence of volunteer selection bias and monetary incentives," *Nature Energy*, 2019, 4, 35–41.

[23] Li, Y., Wen, Y., Tao, D., Guan, K. "Transforming cooling optimization for green data center via deep reinforcement learning," *IEEE Transactions Cybernetics*, 2020, 50, 2002–2013.

[24] Google. Our third decade of climate action: realizing a carbon-free future. 14 September 2020. Available online: https://blog.google/outreach-initiatives/sustainability/our-third-decade-climate-action-realizing-carbon-free-future/ (accessed on 20 December 2022).

[25] Wong, L.A., Ramachandaramurthy, V.K., Walker, S.L., Taylor, P., Sanjari, M.J. "Optimal placement and sizing of battery energy storage system for losses reduction using whale optimization algorithm," *Journal of Energy Storage*, 2019, 26, 100892.

[26] Almulaiki, W.A. "The impact of performance management on employee performance," *Saudi Journal of Business and Management Studies*, 8(2), 2023, pp. 22–27.

[27] Ikonne Chinyere, Prof. Ikonne Chiemela, "Workplace E-monitoring and surveillance of employees: Indirect tool of information gathering," *International Journal of Science and Research (IJSR)*, 3, no. 10, October 2014.

[28] Vilde Hoff Bernstrøm, Helge Svare, "Significance of monitoring and control for employees' felt trust, motivation, and mastery," *Nordic Journal of Working Life Studies*, 7, no. 4, December 2017.

[29] Jeske, D., Kapasi, I. "Electronic performance monitoring: Lessons from the Past and Future Challenges," In *Organizing for digital economy: Societies, communities and individuals. Proceedings of the 14th annual conference of the Italian chapter of the AIS*, 2018, pp. 119–132.

[30] Ahmad, Z., Ong, T.S., Liew, T.H., Norhashim, M. "Security monitoring and information security assurance behaviour among employees: An empirical analysis," *Information & Computer Security*, vol. 27, no. 2, 2019, pp. 165–188.

[31] Vujović V, Maksimović M. "Raspberry Pi as a sensor web node for home automation," *Computers & Electrical Engineering*, 44, 2015, 153–171.

[32] George, A.S., George, A.H., Baskar, T. "The evolution of smart factories: How Industry 5.0 is revolutionizing manufacturing," *Partners Universal Innovative Research Publication*, vol. 1, no. 1, 2023, pp. 33–53.

11 Application of Digital Image Watermarking in the Internet of Things and Machine Learning

K. Prabha and I. Shatheesh Sam

11.1 INTRODUCTION

In an increasingly connected world, the convergence of advanced technologies such as Digital Image Watermarking, the Internet of Things (IoT), and machine learning has opened up new avenues for securing and interacting with digital content. Digital Image watermarking is the technique that embeds invisible data (watermarks) in digital media, such as text, images, audio, and videos, in order to safeguard its ownership, validity, and integrity. This has significant applications in the fields of machine learning and the IoT. This powerful combination helps us to address critical issues such as data security, authentication, and robust image processing.

The IoT, a network of interconnected devices that communicate seamlessly, generates an enormous amount of data every day in this fast growing environment. Frequently, this data consists of sensitive visual information, such as surveillance camera footage, medical images, or industrial sensor data. The incorporation of Digital Image Watermarking provides a robust solution for data protection and authentication. Watermarked images can be reliably traced back to their source, ensuring data integrity and authenticity throughout the IoT ecosystem.

Furthermore, the integration of machine learning techniques enhances the capabilities of Digital Image Watermarking within IoT environments. Machine learning algorithms help in the detection and extraction of watermarks from images, enabling real-time monitoring and analysis. This synergy can prove invaluable in various applications (Hussain et al., 2020), from fraud detection in smart cities to quality control in industrial IoT (IIoT) setups (Shah et al., 2022).

The landscape exploration of Digital Image Watermarking in the IoT and ML will delve into the pivotal role in enhancing data security, ensuring authenticity, and facilitating advanced image processing. This journey will reveal the potential of this combination, shedding light on how it is reshaping the way to safeguard and leverage digital imagery in our increasingly interconnected world.

DOI: 10.1201/9781032623276-11

The contribution of this chapter includes the advantages, disadvantages, and applications of Digital Image Watermarking in IoT and ML. In this chapter, section 2 elaborates the advantages and disadvantages of Digital Image Watermarking in IoT and ML, section 3 illustrates the applications of Digital Image Watermarking in the IoT and machine learning, and section 4 concludes the chapter by stating the importance of digital image watermarking in pointing out the emerging challenges and opportunities in IoT and machine learning.

11.2 ADVANTAGES AND DISADVANTAGES OF DIGITAL IMAGE WATERMARKING IN THE PERSPECTIVE OF BOTH IOT AND ML

There are several advantages of using digital image watermarking in the context of IoT and ML even though it may pose some challenges.

Advantages:

Copyright Protection: Watermarking helps to protect the intellectual property of digital images, ensuring proper attribution, and preventing unauthorized use (Begum & Uddin, 2020).

Authentication: Watermarking can serve as a means of image authentication, image integrity verification as well as tamper protection.

Traceability: Watermarks can embed data about the source or ownership of images, facilitating traceability, and accountability in data usage.

Security in IoT: Watermarking helps to secure communication in IoT applications by confirming the authenticity of images exchanged between devices.

Improved ML Model Robustness: Watermarking can enhance the robustness of ML models by providing additional features or metadata, resulting in better recognition and classification.

Disadvantages:

Quality Degradation: Some watermarking techniques may result in a loss of image quality, impacting the visual appeal and potentially affecting the performance of ML models.

Vulnerability to Attacks: Watermarking may be vulnerable to certain attacks, such as removal or modification, compromising the security and integrity of embedded information.

Computational Overhead: Implementing robust watermarking techniques can introduce computational overhead, especially in resource-constrained IoT devices.

Limited Standardization: The lack of standardized watermarking techniques can lead to interoperability issues, making it challenging for widespread adoption in IoT and ML ecosystems.

Complex Integration in IoT: Integrating watermarking in IoT systems may be complex due to different devices, protocols, and communication channels, requiring careful consideration.

Potential Legal Concerns: Depending upon jurisdiction, the use of digital watermarks may have legal implications, understanding, and complying with relevant laws is essential.

Ethical Considerations: Ethical concerns associated with privacy and consent may arise, especially in IoT applications, where images may be captured and shared without the explicit consent of individuals.

Digital image watermarking brings several advantages, including copyright protection and enhanced security; there are some drawbacks such as potential quality degradation, susceptibility to attacks, and the necessity for careful integration in diverse IoT environments. It is vital to balance these factors and choose appropriate watermarking techniques based on specific use cases and requirements.

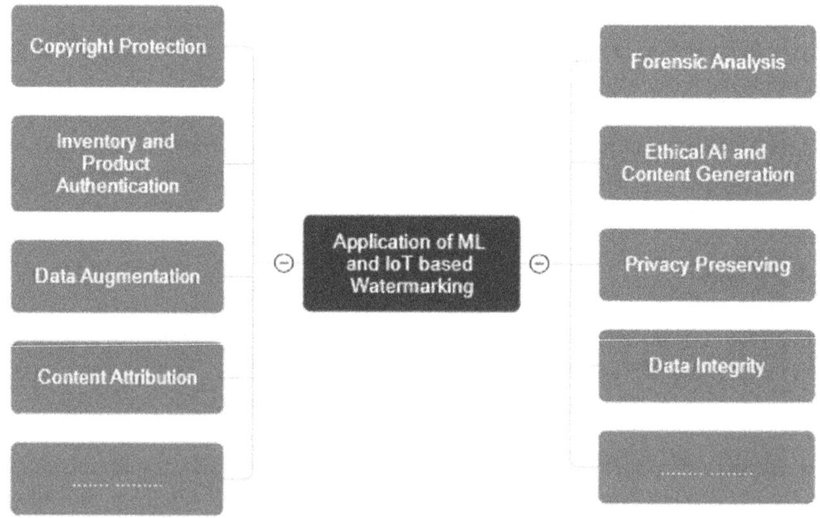

FIGURE 11.1   Applications of ML and IoT-based watermarking.

11.3   APPLICATIONS

When it comes to the IoT and machine learning (ML), digital image watermarking has numerous valuable applications. Figure 1 shows the Applications of ML and IoT based digital image watermarking.

Secure Data Transmission in IoT

Watermarking can embed information like timestamps, device IDs, or encryption keys into images generated by IoT devices. This guarantees the authenticity of the data when transmitted to a central server.

Image Authentication in IoT

In IoT applications such as surveillance cameras, watermarks help to verify the authenticity of captured images or videos (Kamili et al., 2021). This supports tampering prevention with critical visual data.

Copyright Protection

In IoT scenarios where images are shared and exchanged, digital watermarking aids to protect the intellectual property rights of content creators and owners.

Inventory and Product Authentication

Watermarking can be applied to product images in an IoT-based inventory management system. This helps in authenticating products, detecting counterfeit goods, and ensuring data integrity about items in stock.

Medical Imaging and Patient Data

In the context of IoT healthcare applications (Al-kahtani et al., 2022), medical images can be watermarked to maintain data integrity, protect patient privacy, and validate the source of the images.

Data Augmentation in ML

In machine learning models, large datasets are often required for training. Watermarking can be used to create synthetic datasets with known labels, facilitating model training and improving its ability to generalize across different scenarios. For example, watermarking can be applied to images to create synthetic data with established ground truth labels, which can enhance model robustness.

Model Robustness Testing

In ML, watermarking can be used to verify the robustness of image recognition models against common attacks. Watermarked images can be used as part of adversarial testing to evaluate model performance.

Anti-Plagiarism in ML-based Content Generation

In ML, applications such as text-to-image synthesis or image generation, watermarks can be applied to ensure that generated images or content are not plagiarized from external sources.

Content Attribution in Social IoT
In IoT systems that share images and multimedia content on social platforms, digital watermarking can help attribute content to its original creators or sources.

Forensic Analysis
Watermarking can help in the forensic analysis of images and videos collected by IoT devices, helping law enforcement agencies and investigators in solving crimes. It's important to implement robust and secure watermarking techniques to withstand various attacks, including image manipulation and removal attempts. Additionally, compliance with privacy regulations and intellectual property rights is crucial when implementing watermarking in IoT and ML applications.

Improved Data Integrity in IoT
IoT devices often capture and transmit image data in various applications, such as environmental monitoring, agriculture, or industrial automation. By embedding watermarks into these images, data integrity (Zhang et al., 2017) can be maintained, and any unauthorized alterations can be detected. This confirms that decisions made based on IoT data are reliable.

Real-time Authentication in IoT
In real-time IoT scenarios (Behnia et al., 2019), like autonomous vehicles, drones, or remote monitoring systems, image watermarking can serve as a quick and efficient method for verifying the authenticity of images and videos, helping prevent security breaches, and ensuring trust in the data.

Secure Supply Chain Management
In IoT-based supply chain management, watermarking product images or QR codes can be used to authenticate the authenticity of products and their origins. This is particularly useful for combating counterfeiting and ensuring product traceability.

Enhancing Robustness in ML Models
Watermarked images can be used for testing the robustness of ML models against adversarial attacks. This method assures that ML models perform reliably in the environment of potential threats or manipulation attempts.

Ethical AI and Content Generation
In AI applications, watermarking can serve as a tool for ethical AI practices because generated content, such as text or images, is shared publicly. It can attribute AI-generated content to its source, promoting transparency and accountability.

Privacy-Preserving Medical Imaging
In medical IoT, where patient data is shared for diagnosis and treatment, watermarking can be applied to protect patient privacy while allowing data to be exchanged securely among healthcare providers and institutions.

Intellectual Property Protection
Watermarking can safeguard the intellectual property of content creators and artists in IoT applications where images or multimedia content are distributed, shared, or sold.

Anomaly Detection in Security IoT
Watermarking can be applied to surveillance camera feeds in security IoT to detect anomalies or unauthorized access. Any tampering with the watermarked images can trigger alerts for further investigation.

Historical Data Integrity in ML
For historical datasets used in ML research or analysis, watermarking is applied to maintain the integrity of the information and ensure that it remains unaltered over time, preserving the dataset's reliability for future studies.

When implementing digital image watermarking in IoT and machine learning, it's crucial to consider factors such as the robustness of the watermarking technique, computational efficiency, and the precise requirements of the application to ensure its effectiveness. Additionally, staying updated with the latest advancements in watermarking technology and security practices is essential to address evolving threats.

Content Provenance in Social Media and IoT
In the age of social media and user-generated content, digital image watermarking can help establish the origin and ownership of shared images and videos. This is particularly valuable in IoT applications where users contribute multimedia content, ensuring proper attribution and intellectual property protection.

Geo-tagged Image Verification in IoT
For IoT devices with built-in geolocation capabilities, watermarking can include geographic coordinates and timestamps in images. This assists in verifying the authenticity of data gathered from multiple locations, making it useful in environmental monitoring and geospatial applications.

Preventing Data Manipulation in Autonomous Systems
In autonomous systems like self-driving cars and drones, digital image watermarking helps to protect the integrity of sensor data. Any attempts to manipulate sensor-generated images can be detected through watermark verification, enhancing safety and reliability.

Securing Data Marketplaces
In IoT ecosystems where data marketplaces facilitate data exchange, watermarking can be used to ensure the trustworthiness and authenticity of data sources, promoting fair and secure data transactions.

Enhanced Anomaly Detection in ML
Machine learning models, when trained on watermarked datasets containing known anomalies, can be more effective at identifying unusual patterns or outliers in various applications such as fraud detection, cybersecurity, and fault prediction.

Watermarking for Compliance and Auditing
Watermarking IoT-generated images can be essential for compliance with industry regulations and auditing requirements. This affords a mechanism to prove the authenticity of data for regulatory purposes.

Protecting Data Privacy in Surveillance
In IoT-based surveillance systems, watermarks can be applied to protect the privacy of individuals captured in images and videos. Sensitive information, such as faces or license plates, can be obscured with watermarks to comply with privacy regulations.

Digital Evidence in Legal Proceedings
Watermarked images and videos captured by IoT devices can be digital evidence in legal cases. These watermarks act as a tamper-evident seal, ensuring the admissibility and integrity of the evidence in court.

Enhancing User Trust in AI-Generated Content
As AI-generated content becomes more prevalent, watermarking can be employed to indicate that the content is machine-generated, fostering transparency, and trust between users and AI systems.

Securing Remote Sensing Data
In remote sensing applications like satellite imagery and weather monitoring, watermarking can help safeguard data integrity and authenticity, particularly when used for critical decision-making processes.

These applications demonstrate the versatility and importance of digital image watermarking in ensuring data integrity, authenticity, and security in IoT and machine learning contexts. Choosing the right watermarking technique and parameters depends on the specific requirements of each application, including the level of security needed and potential threats that may be encountered.

Data Fusion and Sensor Networks in IoT
In IoT environments with multiple sensors and data sources, watermarking can be used to incorporate data from diverse sensors while ensuring the authenticity and alignment of the information. This is essential for applications like smart cities, where data from various sources, such as traffic cameras and weather stations, are combined.

Adaptive Watermarking in IoT
Adaptive watermarking techniques can be employed in IoT to dynamically adjust watermark strength or presence based on the data's sensitivity. For instance, critical data may have stronger watermarks to protect against tampering, while less sensitive data may have lighter watermarks to save bandwidth.

Visual Search and Content Retrieval in ML
In image-based search and content retrieval systems powered by ML, watermarks can serve as metadata tags, aiding in precise content retrieval and categorization.

Insider Threat Detection in IoT
Watermarking can assist in detecting insider threats within IoT networks. Any unauthorized access or data tampering attempts by individuals with access privileges can be identified through watermark verification.

Protecting Training Data in Federated Learning
In federated learning settings, where ML models are trained across decentralized IoT devices, watermarks can protect the integrity and privacy of local data on these devices. This confirms that sensitive information remains confidential while contributing to the training process.

Anomaly Detection in Medical Imaging
In medical IoT, watermarking can be applied to medical images to detect anomalies, such as tumors or fractures. This assists healthcare professionals in diagnosis and monitoring.

Metadata Preservation in IoT Data Streams
Alongside watermarking, IoT devices can embed essential metadata into images, including device ID, location, and timestamp. Watermarking helps ensure this metadata remains intact throughout data transmission and processing.

Ensuring Fairness in ML Algorithms
Watermarking can add fairness indicators to datasets used in ML algorithms. This guarantees that the training data is well-balanced and represents diverse demographics, mitigating biases in AI systems.

Data Provenance in Environmental Monitoring
In IoT applications related to environmental monitoring, such as climate studies or pollution tracking, watermarking can be employed to establish data provenance, assuring that data gathered from various sensors remains reliable for scientific analysis.

Defense Against Deepfake Attacks
As deepfake technology advances, watermarking plays a vital role in detecting manipulated or synthetic images and videos. Watermarked content can be a reference for authentic media.

Supply Chain Transparency in Agriculture IoT
In precision agriculture IoT, watermarked images can track the entire lifecycle of agricultural products, from planting to distribution. This enhances transparency and accountability in the food supply chain.

Collaborative AI Research
Watermarking is used to protect the intellectual property of research institutions and organizations involved in collaborative AI research and data sharing. It ensures proper attribution and credit for contributions.

Quality Control in Manufacturing IoT
Watermarking can be applied to product images in manufacturing processes monitored by IoT sensors and cameras. This enables real-time quality control by comparing watermarked reference images with the actual product images, ensuring compliance with quality standards.

Verification of Remote Operations in Robotics
In remote robotic operations, such as surgery or exploration, watermarking can be applied to validate the authenticity of video feeds and control signals, safeguarding against unauthorized access or tampering.

Secure Data Sharing in IoT Ecosystems
Watermarking allows for secure data sharing among IoT devices and stakeholders in a networked ecosystem. By embedding watermarks in shared images or data streams, the source and integrity of the data are preserved.

Art Authentication and Provenance
In the art world, digital watermarking can be applied to high-resolution images of artworks to establish their authenticity and provenance. This is crucial for art collectors, galleries, and museums.

Financial Document Verification
In the financial sector, digital watermarking can be applied to secure financial documents, such as checks and certificates. Watermarks can include details like account numbers, dates, and issuer information, enhancing document security.

Historical Data Preservation in Cultural Heritage
Watermarking can be applied to digitized historical documents, photographs, and artifacts in cultural heritage preservation efforts. The integrity of these valuable resources for future generations is maintained by watermark.

IoT-Based Environmental Conservation
In conservation efforts utilizing IoT sensors, watermarked images of wildlife and ecosystems can help track and protect endangered species and habitats, ensuring the data's credibility for research and policy-making.

Personalized Content Generation in ML
ML models can use watermarked images to create personalized content, such as customized advertisements or product recommendations while ensuring the originality and authenticity of the generated content.

Intellectual Property Verification in 3D Printing
In 3D printing applications, digital watermarking (Begum & Uddin, 2020) is used to validate the authenticity of 3D models and designs, protecting the intellectual property of designers and manufacturers.

Agricultural Pest and Disease Monitoring
In the field of precision agriculture, watermarked images are used to detect pests and diseases in addition to monitoring crop health. This data is crucial for producing informed decisions about pesticide use and crop management.

Trustworthy Remote Sensing for Disaster Response
During disaster response efforts that utilize IoT and remote sensing technologies, watermarked images provide reliable and tamper-evident data for disaster assessment, response coordination, and decision-making.

Maintaining Ethical AI in Journalism
Watermarking can be applied to AI-generated content used in journalism to ensure that articles or news reports generated by AI systems are clearly identified as such, maintaining ethical reporting standards.

These diverse applications underscore the significance of digital image watermarking in securing, authenticating, and enhancing data across various IoT and machine learning domains. Implementing the appropriate watermarking techniques depends on specific use cases and the need to balance data security, privacy, and usability.

Augmented Reality (AR) and IoT Integration
In AR applications, IoT devices can capture real-world images enriched with digital overlays. Watermarking can secure these augmented images, guaranteeing that the virtual elements remain reliable in the physical environment.

Watermarking for Privacy-Preserving AI
Watermarking can help preserve privacy in AI research and applications that handle sensitive data. For example, facial recognition models can be generated on watermarked images to avoid unauthorized facial recognition.

Digital Watermarking for Sensor Fusion in IoT
In IoT systems with diverse sensors, watermarks can be used to combine data from different sensors like cameras, accelerometers, and temperature sensors. This creates a comprehensive perspective of the environment while preserving data integrity.

Energy-Efficient Watermarking in IoT
Energy efficiency is crucial in IoT devices. Lightweight watermarking techniques are designed to minimize computational demands, making them suitable for resource-constrained IoT devices.

Weather Forecasting and Climate Research
Watermarked satellite and sensor data are vital for accurate weather forecasting and climate research. These watermarks confirm the authenticity of data used for making critical meteorological predictions.

Enhancing Authentication in ML Model Deployment
When deploying machine learning models in production, watermarked data can enhance user authentication and authorization processes, safeguarding access to sensitive AI models.

Blockchain and Watermarked Data
Watermarked images and data can be incorporated with blockchain technology in IoT and ML applications. This combination ensures an immutable and transparent record of data transactions and model training.

Watermarking for Autonomous Delivery Drones
Drones used in autonomous delivery can capture watermarked images of packages, facilitating tracking, verification, and proof of delivery in logistics operations.

Remote Learning and Watermarked Educational Content
In online education, watermarked content can help verify the authenticity of educational materials. It assures students and institutions that the content is legitimate and from reliable sources.

Watermarking for Product Lifecycle Management (PLM)
In Product Lifecycle Management (PLM) systems used in manufacturing, digital watermarking can be employed to mark computer-aided design (CAD) designs and product images at various stages of development, ensuring design integrity and compliance with specifications.

Watermarking in Recommender Systems
Recommender systems in e-commerce (Sherekar et al., 2008) can use watermarked images and product data to enhance recommendations. This ensures that recommendations are based on authentic product information.

Compliance Monitoring in Healthcare IoT
In healthcare IoT, watermarked patient data and medical images (Hashim et al., 2020) can be monitored for compliance with regulations like Health Insurance Portability and Accountability Act (HIPAA), ensuring patient privacy and data security.

Wildlife Conservation and Watermarked Camera Traps
Camera traps used in wildlife conservation efforts can capture watermarked images of animals, assisting researchers in population studies and habitat conservation.

Watermarking IoT-Generated Art and Creativity

IoT devices can be programmed to generate art or music. Watermarking ensures that such creative works are attributed to their IoT creators and not plagiarized.

These applications illustrate the versatility of digital image watermarking in safeguarding data, enhancing security, and preserving authenticity in both IoT and ML domains. The selection of watermarking technique depends on factors such as the application's specific requirements, data sensitivity, and the prospective for malicious tampering.

Watermarking in Autonomous Vehicles

Autonomous vehicles rely heavily on sensor data, including cameras, LiDAR, and radar. Watermarking can ensure the integrity and authenticity of the data utilized for navigation, obstacle detection, and decision-making, making autonomous transportation safer and more reliable.

Agricultural Yield Prediction

In precision agriculture, watermarked images of crops can be used to forecast yields based on growth patterns, helping farmers to form informed decisions about planting, harvesting, and resource allocation.

Watermarking in Environmental Monitoring Networks

Environmental monitoring networks often employ a multitude of sensors and cameras to observe natural phenomena. Watermarking can assist in integrating and securing data from these sources, ensuring the credibility of environmental research and decision-making.

E-commerce Product Authenticity

In e-commerce (Rahmati et al., 2013), watermarked product images are used to verify the authenticity of products listed for sale. This aids buyers in making informed purchasing decisions while protecting them from counterfeit goods.

Watermarking in Human Activity Recognition (HAR)

Human Activity Recognition (HAR) (Muhammad et al., 2021) applications in IoT and wearables can benefit from watermarking by ensuring the integrity of sensor data used to recognize human activities such as walking, running, or sleeping.

Watermarking IoT Data for Financial Transactions

In IoT-based payment systems and financial transactions, watermarking can be applied to transaction records and receipts to prevent fraud and enhance transaction security.

Watermarking for Authenticating Voice and Sound Data

In addition to images, watermarking can be extended to authenticate voice recordings and sound data in IoT voice assistants, voice-controlled devices, and audio analytics applications.

Watermarking for Wildlife Tracking
In wildlife conservation efforts, watermarked images and videos from camera traps can assist in tracking and studying animal behavior, migration patterns, and population dynamics.

Protecting IoT-Based Digital Twins
In Industry 4.0 and IoT-driven digital twin simulations of physical assets, watermarking is used to secure the precision of the digital representation, ensuring it is arranged with the real-world counterpart.

Watermarking for Preventing Data Poisoning Attacks
In machine learning, watermarking is used to identify data poisoning attacks where adversaries inject malicious data into training sets. Watermarked data can help identify and mitigate such attacks.

Maintaining Data Integrity in Remote Sensing for Agriculture
In precision agriculture, remote sensing data is crucial for crop management. Watermarked data ensures that the imagery used for crop analysis, irrigation planning, and yield prediction remains unaltered and trustworthy.

Geospatial Watermarking for Navigation
In location-based services and GPS navigation applications, watermarked geospatial data can enhance location accuracy and trustworthiness, especially in urban canyons or remote areas.

Watermarking IoT-Generated Artifacts in Industrial IoT (IIoT)
In IIoT settings where machines generate manufacturing instructions, watermarked digital artifacts can ensure the authenticity and traceability of production processes.

These applications emphasize the diverse ways in which digital image watermarking is instrumental in enhancing data security, reliability, and trustworthiness in the evolving fields of IoT and machine learning. Implementing watermarking techniques that align with each application's specific needs and constraints is crucial for their effectiveness.

Watermarking for IoT-Based Environmental Compliance
Industries and municipalities use IoT sensors for environmental compliance monitoring. Watermarked data ensures that air and water quality measurements remain unaltered for regulatory reporting.

Watermarking in Supply Chain Blockchain
In supply chain applications using blockchain, watermarks can be added to product images and sensor data. This helps in verifying the authenticity and provenance of products at each stage of the supply chain.

Watermarking for Sentiment Analysis
In sentiment analysis, watermarked images with associated emotions are applied to train ML models. This enhances the accuracy of sentiment predictions in applications like customer feedback analysis and social media monitoring.

Watermarking for Anti-Counterfeiting in Luxury Goods
In the luxury goods industry, watermarked images of products can be linked to authentication databases, enabling customers to verify the authenticity of their purchases using a smartphone app.

Watermarking for Remote Inspection in Manufacturing
In remote inspection applications, where IoT devices capture images of machinery and infrastructure, watermarked images are used to ensure that scrutiny reports are based on original, unaltered data.

Watermarking for Document Archiving in Legal
Law firms and legal departments can employ watermarking to ensure the integrity of digital documents stored for legal purposes, such as contracts, wills, and evidence.

Watermarking for Smart Grid Security
In the smart grid, watermarked sensor data ensures the reliability and security of the energy distribution system, protecting against unauthorized access or manipulation.

Watermarking for Traffic Management
IoT-based traffic management systems use watermarked images from traffic cameras to analyze congestion, accidents, and road conditions, aiding in real-time traffic control.

Watermarking for Authentication in Augmented Reality Museums
In AR-enhanced museum experiences, watermarked historical images and artifacts can be used to validate the digital overlays, enhancing the educational and immersive aspects of the exhibits.

Watermarking for Cybersecurity Training Data
In cybersecurity training, watermarked datasets containing simulated attacks and vulnerabilities can be used to train ML models for intrusion detection and threat prevention.

Watermarking for Industrial Safety Compliance
In industrial settings, watermarked images can be used for safety compliance inspections, ensuring that equipment inspections and safety protocols are based on unaltered data.

Watermarking for IoT-Based Wildlife Conservation Databases
Conservation organizations can use watermarked images and data to maintain reliable databases for tracking wildlife populations and protecting endangered species.

Watermarking in Smart Grid Meter Reading
In smart grid meter readings, watermarked meter images are used to authenticate the accuracy of energy consumption data, reducing billing disputes and fraud.

Watermarking for Personal Health Monitoring
In wearable health devices, watermarked health data are used for personal health monitoring and medical diagnoses, ensuring data integrity and privacy.

Watermarking for Drone-Based Infrastructure Inspection
Cameras Drones can capture watermarked images of critical infrastructure such as bridges and power lines, ensuring the integrity of inspection reports.

These diverse applications demonstrate the critical role that digital image watermarking plays in safeguarding data, enhancing authenticity, and ensuring the trustworthiness of information across various sectors, from environmental monitoring to cybersecurity and cultural preservation. Tailoring watermarking techniques to the specific requirements of each use case is essential for achieving optimal results.

Data Deduplication and Watermarking
In IoT, where immense amounts of data are generated, watermarking is used to identify duplicate data entries, aiding in data storage efficiency and ensuring that redundant information is properly labeled.

Watermarking in Edge Computing
In edge computing environments, watermarked data can be processed and verified locally, reducing the necessity for extensive data transmission and enhancing data security in IoT applications.

Privacy-Preserving Collaborative ML
Watermarking can be incorporated into federated learning and collaborative ML approaches, allowing multiple parties to train models without sharing sensitive raw data. Watermarks can attest to the credibility of contributed data.

Watermarking for Cross-Domain Image Retrieval
In ML-based image retrieval systems, watermarked images are used to improve cross-domain retrieval accuracy by preserving important metadata and context information.

Watermarking for Image Forensics
Watermarking techniques can be instrumental in image forensics, enabling investigators to identify tampered or manipulated images in legal and criminal investigations.

Watermarking for Resource-Constrained IoT Devices
Lightweight watermarking algorithms are developed to be efficient and suitable for resource-constrained IoT devices with limited memory and processing power.

Watermarking for Secure Remote Voting Systems
In remote voting systems, watermarked ballots can help ensure the integrity of votes and prevent tampering, thus enhancing the security and transparency of the voting process.

Explainable AI and Watermarking
Watermarked training data is used to improve the interpretability of machine learning models, contributing to explainable AI by preserving information about critical features and input variables.

Watermarking for Digital Preservation
In digital libraries and archives, watermarked images and documents can serve as digital artifacts, preserving cultural heritage and historical records while confirming their authenticity.

Blockchain-Enabled Watermarking
Combining watermarking with blockchain technology (Wang & Hsu, 2022) offers a robust solution for data integrity and provenance tracking in diverse applications, comprising supply chain management and legal documentation.

Privacy-Preserving AI Services
Watermarking can be applied to data used by AI services to confirm that user data remains confidential and untouched during model training and inference, supporting privacy-preserving AI.

Watermarking for Monitoring Wildlife Behavior
In wildlife behavior research, watermarked images are used to monitor and track animal movements, interactions, and habits without disturbing their natural environment.

Watermarking for Precision Medicine
In personalized medicine, watermarked medical images (Allaf & Kbir, 2019) and genetic data can facilitate the development of tailored treatments while maintaining patient data security and authenticity.

Continuous Monitoring of Industrial Equipment
Watermarked images and sensor data from industrial equipment enable continuous monitoring, ensuring that machinery operates within safe and efficient parameters and alerting to any anomalies.

11.4 CONCLUSION

Digital Image Watermarking emerges as a vital tool in this context, providing robust mechanisms for data protection, authentication, and ensuring the integrity of visual content. With its ability to process and analyze vast datasets, machine learning complements Digital Image Watermarking by enhancing the detection and extraction of watermarks from images. This synergy opens up new avenues for real-time monitoring and analysis, fostering innovation in various domains. These advanced applications underscore the adaptability and importance of digital image watermarking in addressing evolving challenges and opportunities in IoT and machine learning. Watermarking techniques should be chosen carefully, considering the specific needs and constraints of each use case, to achieve optimal results in terms of data security, authenticity, and trustworthiness. In summary, digital image watermarking finds extensive applications in IoT and ML, ranging from data integrity and privacy protection to content attribution and fairness enhancement.

REFERENCES

[1] Al-kahtani, M. S., Khan, F., & Taekeun, W. (2022). Application of Internet of Things and Sensors in Healthcare. *Sensors*, *22*(15), 5738. https://doi.org/10.3390/s22155738

[2] Allaf, A. H., & Kbir, M. A. (2019). A Review of Digital Watermarking Applications for Medical Image Exchange Security. In M. Ben Ahmed, A. A. Boudhir, & A. Younes (Eds.), *Innovations in Smart Cities Applications Edition 2* (pp. 472–480). Springer International Publishing. https://doi.org/10.1007/978-3-030-11196-0_40

[3] Begum, M., & Uddin, M. S. (2020). Digital Image Watermarking Techniques: A Review. *Information*, *11*(2), 110. https://doi.org/10.3390/info11020110

[4] Behnia, R., Ozmen, M. O., & Yavuz, A. A. (2019). ARIS: Authentication for Real-Time IoT Systems. *ICC 2019 – 2019 IEEE International Conference on Communications (ICC)*, 1–6. https://doi.org/10.1109/ICC.2019.8761207

[5] Hashim, M. M., Rhaif, S. H., Abdulrazzaq, A. A., Ali, A. H., & Taha, M. S. (2020). Based on IoT Healthcare Application for Medical Data Authentication: Towards A New Secure Framework Using Steganography. *IOP Conference Series: Materials Science and Engineering*, *881*(1), 012120. https://doi.org/10.1088/1757-899X/881/1/012120

[6] Hussain, F., Hassan, S. A., Hussain, R., & Hossain, E. (2020). Machine Learning for Resource Management in Cellular and IoT Networks: Potentials, Current Solutions, and Open Challenges. *IEEE Communications Surveys & Tutorials*, *22*(2), 1251–1275. https://doi.org/10.1109/COMST.2020.2964534

[7] Kamili, A., Hurrah, N. N., Parah, S. A., Bhat, G. M., & Muhammad, K. (2021). DWFCAT: Dual Watermarking Framework for Industrial Image Authentication and Tamper Localization. *IEEE Transactions on Industrial Informatics*, *17*(7), 5108–5117. https://doi.org/10.1109/TII.2020.3028612

[8] Muhammad, K., Mustaqeem, Ullah, A., Imran, A. S., Sajjad, M., Kiran, M. S., Sannino, G., & De Albuquerque, V. H. C. (2021). Human Action Recognition Using Attention Based LSTM Network with Dilated CNN Features. *Future Generation Computer Systems*, *125*, 820–830. https://doi.org/10.1016/j.future.2021.06.045

[9] Rahmati, P., Adler, A., & Tran, T. (2013). Watermarking in E-commerce. *International Journal of Advanced Computer Science and Applications*, *4*(6), 256–306. https://doi.org/10.14569/IJACSA.2013.040634

[10] Shah, S., Iqbal, M., Aziz, Z., Rana, T., Khalid, A., Cheah, Y.-N., & Arif, M. (2022). The Role of Machine Learning and the Internet of Things in Smart Buildings for Energy Efficiency. *Applied Sciences*, *12*(15), 7882. https://doi.org/10.3390/app12157882

[11] Sherekar, M. S. S., Thakare, D. V. M., & Jain, D. S. (2008). Role of Digital Watermark in E-governance and E-commerce. *International Journal of Computer Science and Network Security*, *8*(1), 257–261.

[12] Wang, C.-H., & Hsu, C.-H. (2022). Blockchain of Resource-Efficient Anonymity Protection with Watermarking for IoT Big Data Market. *Cryptography*, *6*(4), 49. https://doi.org/10.3390/cryptography6040049

[13] Zhang, G., Kou, L., Zhang, L., Liu, C., Da, Q., & Sun, J. (2017). A New Digital Watermarking Method for Data Integrity Protection in the Perception Layer of IoT. *Security and Communication Networks*, *2017*, 1–12. https://doi.org/10.1155/2017/3126010

Index

A

Actuators, 21, 24, 45
Augmented reality, 187

B

Bigdata, 33, 146
Blockchain, 97, 174
Bluetooth, 58, 134

C

Cellular networks, 59
Clone block attack, 97
Cloud computing, 22, 33
Cobots, 174
Context-dependent sentiment classification, 145
Convolutional neural network, 145
Cutting edge technologies, 130
Cybersecurity, 5

D

Database as a Service (DBaaS), 47
Data heterogeneity, 70
Decision trees, 71
Digital watermarking, 187

E

Edge computing, 22, 35, 59
Embedded systems, 25
Encircling, 104
Ensemble learning, 4

F

Feedforward neural network, 145
Field programmable gate array, 111
5G Connectivity, 33

G

Global roaming, 60
Grey wolf optimization, 103

H

Honey badger optimization, 113

I

Industry 4.0, 190
Instance-based learning, 5
Intelligent IoT, 33
Intelligent transportation system, 6
Internet of Things, 127
Internet of Things in industry, 28, 33
IoET, 174

K

K-Means clustering, 74

L

Linear regression, 153
Long Term Evolution Advanced (LTE-A), 135
Long-Term Evolution (LTE), 135
Low-power WAN, 58

M

Machine-to-Machine (M2M) communication, 33
Mesh networking, 59
Microcontrollers, 24, 45
Mobile edge computing, 97
MQTT, 166
MultinomialNB, 146
Multipliers, 112
Multitask learning, 4

N

NB-IoT, 58
Network functions virtualization, 136
Neural network learning, 4
NLP, 113

P

Perception layer, 40
Principal component analysis, 145

Q

Quantum key distribution, 136
Quantum networking, 136

R

Random forest, 73
Reinforcement learning, 3, 82
Renewable energy, 7
RFID, 33, 133, 164

S

Semi-supervised learning, 3
Sensing and Actuation as a Service (SAaaS),
    47
Sensing as a Service (SaaS), 47
Sensor as a Service (SenaaS), 47
Sensors, 21, 24, 45
Sigfox, 58
6LoWPAN, 43
Smart grid, 130
Supervised learning, 2, 81
Support vector machines, 71, 115

T

Three-layer architecture, 40
Title firmness, 42
Transfer learning, 105

U

Ultra-wideband, 134
Unsupervised learning, 2, 81

V

VGG16, 107
VGG16-TL, 102

W

Wearable devices, 27
Wireless fidelity, 134
Wireless sensor networks, 134

Z

ZigBee, 42, 58, 134
Z-wave, 58

For Product Safety Concerns and Information please contact our EU
representative GPSR@taylorandfrancis.com
Taylor & Francis Verlag GmbH, Kaufingerstraße 24, 80331 München, Germany

www.ingramcontent.com/pod-product-compliance
Ingram Content Group UK Ltd.
Pitfield, Milton Keynes, MK11 3LW, UK
UKHW022315100726
473146UK00009B/479